I0462003

Great British Buses by John A Godwin

© John A Godwin 2012-2020

First published in electronic format in May 2012.
3rd Edition published in January 2020.

Dewey Decimal Classification: 388.3

ISBN: 978-0-244-84778-4

Visit www.buspix.uk to view our online collection of buses, coaches, trolleybuses and trams from more than 50 countries around the world. Join in the conversation on our Facebook page too.

Introduction

For over 100 years, British manufacturers have been busy designing, manufacturing and improving a wide range of passenger vehicles, from open platform double deck buses to the latest in environmentally friendly hybrids. Famous names such as AEC, Bedford, Bristol, Daimler and Leyland may no longer be with us, being victims of recession, competition or international acquisitions, but their legacy lives on in many of the vehicles on Britain's roads today. Modern British vehicles may carry the name of Alexander Dennis, Optare or Wrightbus, reflecting a revived and growing British bus industry that is responding to increasing passenger numbers and operator demand for modern, low-floor and environmentally friendly vehicles.

What makes a "great" British bus? With hundreds of eligible designs to choose from over the last century I have selected fifty which in my opinion define the British bus, whether they introduced innovative designs or technological advances, filled the order books, or are simply the buses best remembered by the past and present travelling public. By British, I have endeavoured to select from those built by British companies within the United Kingdom, although this definition may be challenged in one or two cases. For example, in recent years Optare sold their majority shareholding to the Indian Ashok Leyland company, although I have decided that the Optare Solo and Versa are worthy of retaining their places here as an example of the ongoing contribution being made by Optare to the current British bus scene.

It's a challenge to summarise a particular bus model in just a page or two, when many have had complete books dedicated to them in their own right. However, I have endeavoured to provide an accurate, informative and concise history of each vehicle, from its origins to its demise, highlighting its development over the intervening years, and identifying some of the key operators who used it. Each vehicle is illustrated with photographs, helping to revive memories of decades past as well as today's modern fleets.

John A Godwin, Hampshire
January 2020

Contents

Guy Arab (1933-1969)

Guy Motors of Wolverhampton in the West Midlands has been producing truck chassis which had also been used for buses since 1914 and introduced a specific bus chassis range in 1925. The Guy Arab chassis was introduced in 1933, and was made available as a single deck, double deck or six-wheeled version. It was advertised as the first bus chassis to be designed specifically for the diesel engine, at a time when many manufacturers were still using petrol engines, and Guy had developed a strong relationship with engine manufacturer Gardner. The single deck Arab was offered with the four-cylinder 4LW or five-cylinder 5LW engine, whilst the double deck version had the 5LW or larger six-cylinder 6LW fitted.

There was a reduction in Guy Arab production until 1942, due to Guy being engaged in the production of military vehicles. This is the reason why they were asked to produce a wartime version of the Guy Arab, which was widely adopted by many fleets during the Second World War. This was a heavy chassis made with cast iron components replacing aluminium, which was needed for the war effort. The reliability of the Guy Arab I and longer-nosed Guy Arab II brought about much post-war interest from operators, and a new single deck Guy Arab was introduced in 1946. High demand saw Guy starting to produce its own single deck bodies for the Arab, and in 1948 reached an agreement with Park Royal Vehicles of London to provide double deck bodies.

A revised Guy Arab III was soon added to the range, which provided a batch of 20 Guy-bodied examples to BMMO in 1949, these "GD6" class vehicles being unusual in having the more powerful Meadows diesel engine fitted. A new Mark IV version followed, as a direct result of Guy receiving a 100-vehicle order from Birmingham Corporation in 1950. The Mark IV was available as a 27' or 30' chassis, with a choice of Garner 5LW or 6LW engines although the 5LW was not fitted to the longer chassis as it was not considered powerful enough for the heavier bus. It was visibly different to previous Arabs due to its

Above: Southdown Motor Services were a significant operator of the Guy Arab in the 1940s and 50s. 547 (PUF647) is a Guy Arab IV from 1956, fitted with a Park Royal 59 sear body and uniquely having rear platform doors. 547 was withdrawn by Southdown in 1970.

modernised front radiator grille. The Mark IV soon lead to the demise of the short-lived Mark III in 1953, although an underfloor-engined Arab ("Arab UF") and lightweight version ("Arab LUF") were also introduced and produced for most of the 1950s.

The Guy Arab IV continued to be manufactured alongside the new Guy Wulfrunian, which was introduced in 1958. Following growing financial difficulties, Guy Motors became insolvent in October 1961 and was soon rescued by Jaguar Motors, becoming Guy Motors (Europe) Ltd. The final Arab, the Mark V, was introduced in 1962 and included full air brakes, a lower frame, and telescopic front and rear shock absorbers. It was fitted with a Gardner 6LW engine with four-speed gearbox. With the increasing interest in rear-engined buses and fleet modernisation programmes driven by the Government's bus grant, the Guy Arab was retired from production in 1969, the last examples being delivered to Chester Corporation. Guy Motors ceased production of all models in 1986, after the Guy Victory export model was finally withdrawn.

Bristol K (1936-1948)

Bristol had been producing their G-type double deck bus chassis with petrol or diesel engines since 1931. The Bristol K was developed as a replacement for these but was only offered with a diesel engine from AEC, Bristol (six-cylinder) or Gardner (four, five or six-cylinder). First produced in 1936, it was not until after the end of the Second World War that the chassis was upgraded: changes to Construction and Use Regulations in 1950 meant that the length could be extended to 27'6" (becoming the KS, the "S" indicating the longer chassis) and also take bodywork up to 8' wide (becoming the KSW, the "W" indicating the increased width). Bodywork was manufactured by Eastern Coach Works at Lowestoft and was available in two heights with an open rear platform. All Bristol K types had exposed front radiators, to which the scripted "Bristol" badge was affixed in the top left-hand corner.

The control of the supply of buses by the Ministry of War Transport between 1939 and 1945 saw London Transport taking several Bristol Ks, a type not normally associated with the Capital. Nine B-class Bristols were finished by Park Royal with wartime utility bodywork in 1942, becoming B1-B9 (FXT419-427). A second batch were ordered in 1943, these being specified with AEC engines and Duple bodywork, and B10-B29 (HGC235-254) arrived after the end of hostilities in 1945. The B-class were soon ousted by new AEC Regent III RTs by 1953, many being rebodied by ECW before passing to Brighton, Hove & District, Crosville, Lincolnshire and United (for Hartlepool Borough Council). London Transport found itself with more Bristol Ks for a brief spell in 1951, following the takeover of the Eastern National bus network in the Essex town of Grays, although they were all returned as standard London stock was drafted in towards the end of that year.

Whilst Bristol had supplied K chassis to a variety of operators since its launch, the formation of the British Transport Commission in 1948 included the former Tilling Group bus companies, and as such Bristol were limited to providing vehicles only to BTC fleets from that point onwards. Sizeable fleets were created by many companies, including Brighton, Hove & District, Bristol Omnibus, Bristol Tramways, Eastern National, Lincolnshire Road Car, Southern Vectis, Thames Valley and West Yorkshire Road Car, to name but a few.

Bristol started to develop the new Lodekka in 1949, and the commencement of full Lodekka production in 1954 started a decline in Bristol K orders and production. The last examples were a batch of eight Gardner powered KS6Gs for Brighton, Hove & District in 1957, MPM493-500 (493-500) having 62-seat ECW bodywork with 35 on the upper deck and 27 on the lower. They were specified as only 7'6" wide and were intended for services operating on the narrow two-way St James Street. All surviving Brighton, Hove & District Bristol Ks passed to Southdown Motor Services upon the creation of the National Bus Company in 1969, and only had short lives being scrapped soon after.

One Brighton, Hove & District KSW6G does survive: 1953 HAP985 is now actively preserved by Go-Ahead subsidiary company Brighton & Hove and is occasionally used on special services and for private hires. Another preserved Bristol K worthy of note is Southern Vectis CDL899, which was delivered to the Isle of Wight in July 1939 and is still owned by its original operator today (as Go South Coast Ltd). Converted to open top format in 1958, CDL899 continued to provide occasional public service until 2006, but in recent years its public outings have been less numerous although it can still normally be seen visiting the Epsom Derby on the first Saturday of June each year. Similar 1940 DDL50 also remains on the island, in the care of the Isle of Wight Bus Museum at Newport.

Above: HAP985 is a 1954 Bristol KSW6G, with 60 seat bodywork supplied by Eastern Coach Works at Lowestoft. New to Brighton, Hove & District, it worked until 1968 and was taken into the fleet of Southdown Motor Services upon formation of the NBC 1969.

AEC Regent III (1938-1956) & V (1954-1968)

The Associated Equipment Company of Southall introduced the AEC Regent III in the late 1930s. Two distinct versions were constructed, the first being the "RT" version developed in conjunction with London Transport for use in the Capital from 1938, and the second being a post-war standard version designed to replace the petrol-engined AEC Regent II in 1947, and which was made available to operators outside of London as well as export customers. The Regent III was powered by a 9.6-litre AEC engine, driven by a pre-select or crash gearbox, and was fitted with air brakes.

The London RT version was built specifically for use in the Capital, and less than 100 were built for other operators. LT produced jigs for the RT's bodywork, which were then supplied to Park Royal and Weymann to manufacture the bodies: 150 buses being ordered in 1939. After the Second World War, AEC produced the RT chassis quicker than Park Royal and Weymann could complete them, so additional capacity, albeit to a non-standard design, was provided by Saunders of Anglesey and Craven of Sheffield. The situation reversed a short while later with bodywork production exceeding chassis availability, and in 1949 LT decided to place Park Royal and Metro Cammell produced bodies on Leyland PD2 7'6" wide chassis (becoming the RTL class). Leyland also built 500 of their own "RT" bodies onto wider 8'0" Leyland PD2 chassis (becoming the RTW class). The combined London Transport "RT Family" numbered 6,956 comprising 4,825 RTs, 1,631 RTLs and 500 RTWs by the final delivery in 1954. The RT continued in use in the Capital until RT624 (JXC432) performed the last run on route 62 from Barking Garage on 7th April 1979.

For non-London fleets, bodywork for the post-war Regent III was supplied in both normal and lowbridge formats by companies including East Lancashire, Metro Cammell, Park Royal, Roe, Strachan and Weymann. The Regent III was a very popular double deck vehicle throughout the 1950s and 60s, being used by most bus operators, and the last Regent III (MRD147) was a lowbridge Park Royal bodied example delivered to Reading Corporation as fleet number 4 in 1956. London Transport found itself in need of new low height double deck buses in the early 1950s, and ordered 76 RLH (Regent Low Height) buses with a reduced overall height of only 13'4", this being achieved by the use of sunken offside gangway on the upper deck, and seats in rows of four. 26 were painted red for Central Area services with low bridges, whilst 50 were green for similar Country Area services.

Above: AEC Regent III RT190 (HLW177) was delivered new to London Transport in October 1947 being allocated to Leyton. During overhaul in May 1957, its Park Royal body ended up on the chassis of 1949 RT1173 (JXC481). It served until withdrawal in 1963.

Unusually, between 1950 and 1957 138 AEC Regent III bus chassis were completed with fire appliance bodywork by Maudslay Motors of Alcester before being supplied to fire brigades throughout the UK.

The successor to the Regent III, the Regent V, commenced production in 1954, and was identifiable by its modernised frontal styling, including the radiator being located behind the front grille. The Regent V was powered by either an AEC or Gardner 6LW engine, which was connected to an AEC Monocontrol semi-automatic or fully automatic gearbox.

Some were also fitted with the increasingly popular forward door format instead of a conventional rear platform. East Kent Road Car were a significant user of the Regent V, taking delivery of their first 40 in 1959 fitted with distinctive full front Park Royal bodywork for 72 passengers and a forward entrance. The initial batch (PFN843-882) was known locally as "Puffins" due to their registration numbers, and East Kent placed significant additional orders on almost an annual basis, including the WFN batch in 1961, YJG in 1962, FN in 1963, AFN-B in 1964, GJG-D in 1966 and finally MFN-F in 1967. Whilst the Regent V was never used by London Transport, one of the East Kent buses (GJG750D) passed to Leaside Buses in 1991 for use on school journeys, as a driver trainer and for occasional private hire. Given the appropriate fleet number "RV1", it was never used in normal passenger service.

As with earlier Regent models, significant numbers were built for overseas operators. Left hand drive Regent Vs were supplied in large batches to Lisbon Electric Tramways between 1957 and 1961 (bodied in the UK by Weymann or locally by CCFL or UTIC), SAVR in Teheran (Iran) between 1958 and 1967 (Park Royal bodied) and PTS in Baghdad (Iraq) between 1958 and 1966 (also Park Royal bodied). Right hand drive models were supplied to Johannesburg MT in South Africa between 1956 and 1959 (local bodied by Busaf), West Pakistan Road Transport Corporation between 1961 and 1967 (bodied in the UK by MCCW or Weymann, or supplied as chassis only) and the largest export order of 210 supplied to Kowloon Motor Bus in Hong Kong between 1963 and 1966, bodied by BACo or using Metal Section kits), these being the longest Regent V chassis ever built at 34'.

The availability of the 1968 Government Bus Grant, which was implemented to encourage bus fleet modernisation and the adoption of one-man operation, effectively ended demand for the front-engined AEC Regent V. The last completed buses were 409/410 LMN for Douglas Corporation in December 1968, with 64-seat bodies built by Willowbrook.

Above: Eastbourne Corporation operated many Regents: its final batch of Regent Vs was delivered in 1963. 69 (KHC369) had a 9.6 litre AEC engine and East Lancs bodywork for 60 seated passengers. Platform doors were fitted in 1971 for use on private hire duties.

London Transport's RTs have had roles in a few major films since their withdrawal. Three former London Transport Regent RTs became one "RT1881" (WLB991) for Cliff Richard's 1962 "Summer Holiday" film, these being RT2305 (KGU334), RT2366 (KGU 395) and RT4326 (NLE990), although one of these never left the UK. The 1973 James Bond film "Live and Let Die" saw RT246 (HLW233) purchased by Pinewood Studios and sent to Jamaica for the well-known low bridge chase scene. More recently, three RTs were used to make the triple deck "Knight Bus" in the 2004 film "Harry Potter and the Prisoner of Azkaban", these being RT2240 (KGU169), RT3882 (LLU681) and RT4497 (OLD717).

Bedford OB & OWB (1939-1951)

Bedford introduced their new "O" goods chassis in the late 1930s, and in 1939 commenced production of the "OB" coach as a replacement for the existing Bedford WTB. Costing £850, these vehicles with their characteristic bull-nose grille were bodied by Duple using their Hendonian design on the 14'6" wheelbase chassis, and were fitted with Bedford's 3519cc, six-cylinder petrol OHV engine, coupled to a four-speed crash gearbox. The engine and gearbox were mounted at a slight angle, which combined with an offset differential on the rear axle allowed for a sunken gangway for easier passenger access. The pre-war period saw only 73 examples completed before production was diverted to support the war effort. The Ministry of Supply later authorised Bedford to produce a wartime version of the OB (the Bedford OWB), which was constructed upon a wooden frame and only designed to have a limited life span. Many OWBs were delivered in wartime grey, and the basic specification reduced passenger comfort with the installation of slatted wooden seats. More than 3,300 are known to have been produced, bodied by Duple and Roe.

Bedford OB production re-commenced in 1946, almost exclusively using Duple's new Vista bodywork design. The Vista bodywork was constructed on a wooden framework made from ash reinforced with steel and covered with aluminium panels, was fitted with 27 or 29 coach seats and overhead luggage racks and incorporated a rear boot which also housed the spare wheel. A sightseeing version was available which included overhead roof glazing and modified luggage racks. The complete vehicle was offered to operators at £1,314 for the 27-seat version, and only £1,325 for 29 seats. A number of wartime OWB chassis were re-bodied with the Duple Vista body as a means of bypassing the long post-war waiting list for new Bedford OB chassis. Several OBs were also bodied by other companies, including Mulliner, Plaxton, Strachan and Thurgood.

The Bedford OB was popular with operators of all sizes, with examples working for larger fleets such as Crosville, Grey Cars, Hants & Sussex, King Alfred of Winchester, Southern National and Southern Vectis. Southdown acquired two in 1948 to operate a bus service to Hayling

Island across a weak bridge, and the OB's width made it a popular choice for Jersey Motor Transport and several coach tour operators using Jersey's narrow and winding roads. More unusual operators included British Railways, BOAC (for airport work) and the Metropolitan Police Service who received a fleet of 25 in May 1949: these OB buses were not fitted with internal luggage racks were based at Henley for police officer and prisoner transport duties until the mid-1960s.

Above: BJV590 is a Bedford OB with Duple's Vista bodywork, which was new to Granville Tours of Blackbourn near Grimsby in August 1950. It subsequently served with Kestrel Coaches of Worcester and then became Callander's "Trossachs Trundler" in Scotland between 1993 and 1999. Until recently it was owned by Barnes of Swindon.

Bedford introduced the new larger "SB" chassis at the 1950 Commercial Motor Show, and Bedford OB production ended shortly afterwards in 1951 after a production run of 12,693 vehicles. Unfortunately, many operators found the SB to be too large for their needs, and they had to wait until 1961 for the introduction of the smaller Bedford VAS chassis to fill the gap, which was suitable for 29-seat bodywork. To this day Bedford OBs continue to be operated as vintage vehicles by a small number of UK operators, and they enjoy a healthy level of support from preservationists, being regular visitors to rallies and events all over the country. A 70th anniversary event was held in 2009 at Vauxhall's premises in Luton which 30 Bedford OBs attended. They can regularly be seen in televised period dramas including Foyle's War and Miss Marple.

Crossley DD42 (1944-1952)

During the Second World War, the production of buses and coaches was limited to a few authorised "utility" models, but this did allow some companies the opportunity to plan their new post-war designs. Crossley Motors of Manchester had been working on a new direct injection engine since 1938, and by 1942 several the new 8.6-litre "HOE7" engines had been completed. Fitted to two new chassis, one carrying the body of a Manchester Corporation Crossley Mancunian (GNE217) the "DD42" (denoting Double Deck, 1942) were successful in trials and the government authorised 150 double deck chassis and 80 bodies to be produced in 1944. A single deck version, the SD42, was also being designed, but this version had to wait until 1947 to enter production.

Production commenced in 1945, with the majority of DD42s receiving 56 seat, rear entrance, half cab bodywork that had been constructed by Crossley itself, with a lowbridge option. In 1946, Crossley relocated production to a former government facility at Erwood Park in Stockport and the previous premises at Gorton were closed and sold. Until 1950, the "Manchester Corporation influence" lead to the Crossley bodied DD42 being built with Metro Cammell frames, although the 1948 takeover by AEC lead to these being replaced with Park Royal frames. Some operators chose to have their DD42 chassis completed with other bodywork, including Brush, East Lancashire, Massey, Metro Cammell,

Reading, Roe or Scottish Commercial. The DD42 was powered by Crossley's own 8.6-litre HOE7 diesel engine, connected to a four-speed transmission, and were economical in returning fuel economy of over 10mpg. The chassis was available in 7'6" and 8'0" widths and was fitted with front and rear spring suspension and vacuum assisted brakes. The chassis alone cost £1,630, which was relatively expensive for a bus at the time.

Early deliveries of DD42s were made to Bury (5), Luton (8), Manchester (71) and South Shields (6). Other operators placing orders included Bradford, Chesterfield, Colchester, Derby, Lancaster, Leeds, Plymouth, Portsmouth, Reading, Rotherham, Stockport and Sunderland. Unusually, Luton, Plymouth and Reading Corporations were the only municipals who requested the lowbridge option. Ultimately, the two largest operators were Manchester Corporation with 291 examples, closely followed by Birmingham Corporation with 270. In 1945, Crossley received a substantial export order from Netherlands Railways: amongst their requirements were 250 tractor units which were to be used for articulated trailer buses. The DD42 chassis was specially shortened for this purpose, with the passenger trailers being locally built in the Netherlands for a maximum of 80 passengers.

Crossley DD42 Specifications

- Prototypes – HOE7 engine (1944)
- DD42/3 – HOE7/1 engine, 7'6" wide (1945-47)
- DD42/4 – HOE7/1 engine, 8'0" wide (1947-48)
- DD42/5 – HOE7/1 engine, 7'6" wide (1947-48)
- DD42/6 – HOE7/4, 7/4B, 7/5B engine, 7'6" wide (1948-50)
- DD42/7 – HOE7/4, 7/4B, 7/5B engine, 7'6" wide (1948-50)
- DD42/8 – HOE7/4, 7/4B, 7/5B engine, 8'0" wide (1948-50)
- DD42/7 – HOE7/5 engine, 7'6" wide (1950-52)
- DD42/8 – HOE7/5 engine, 8'0" wide (1950-52)
- DD42/8A – HOE7/5 engine, 8'0" wide (1950-52)

Above: GR9007 is one of twelve Crossley DD42/3s delivered to Sunderland Corporation Transport between 1947 and 1949, six of which were fitted with Crossley bodywork. The original Crossley diesel engines were replaced with Gardner 5LW units from pre-war Daimler buses. This example was withdrawn from service in 1962.

After seeking potential business partners, Crossley was sold to AEC in 1948 as part of the Associated Commercial Vehicles (ACV) group. Crossley's high-end prices and a shortage of raw materials affected orders, and it was decided that chassis production would formally end in 1951.

The last Crossley DD42s were completed in 1952, and amongst the very last orders were a batch of Brush bodied DD42/8As for Derby Corporation. A total of 1,114 Crossley DD42s had been produced in its seven-year production run.

Leyland Titan PD2 & PD3 (1946-1970)

Leyland has been producing the Leyland Titan TD-series of double deck bus chassis since 1927, and production was halted in 1942 due to the Second World War. Once hostilities were over, it was not until 1946 that Leyland resumed chassis production, and these were to be high capacity vehicles to meet operators' requirements. The post-war chassis was a completely new design and known as the Leyland Titan PD1, and interestingly shared major components with the new Leyland Tiger PS1. The first Titan PD1 was EN8536, a 56-seat Roe bodied example, delivered in March 1946 to Bury Corporation.

A year later in 1947, the Leyland Titan PD2/1 was launched, using the Leyland O.600 engine that had been used on lorry chassis since 1946. This was a much-improved engine requiring less servicing and offering an improved vacuum braking system. The PD2/1 was later made available at 8" wide, becoming the PD2/3. In 1948, London Transport ordered 2,135 PD2 Titans to assist in the London fleet replacement programme as AEC could not produce their Regent III chassis fast enough. 7'6" wide PD2/1 with 56-seat Park Royal bodywork became their "RTL" class, whilst the wider 8'0" PD2/3 chassis became their "RTW" class.

In 1950, the Construction and Use Regulations allowed the maximum length for double deck buses to be increased to 27'0", Leyland taking advantage of this to extend the PD2, with 7'6" wide models known as PD2/10 and the 8" models PD2/12. The extended length allowed seating capacity to be extended from 56 to 58, although some body manufacturers managed to raise this to over 60. A further length extension to 30'0" followed in 1956, with Leyland producing a new range of Leyland Titans with the designation "PD3": six different versions were made available (PD3/1-PD3/6) and all were to the maximum 8' width. This additional length also allowed Leyland to commence planning for the rear-engined Leyland Atlantean, which was soon to replace the Titan.

Above: Maidstone Corporation Transport ordered several batches of Leyland PD2s with Massey rear-entrance bodywork between 1956 and 1963. 26YKO was amongst the last and retained its original brown and cream livery until repaint into Maidstone's blue and cream scheme in August 1966, regaining original livery much later in 1983.

The Leyland Titan was a popular and widely adopted double deck chassis, both in the UK and for export orders, and was completed with bodywork by almost every bodybuilder at the time. With so many chassis and body combinations available, certain configurations became associated with specific fleets, for example the Leyland PD3/4 with full-front Northern Counties bodywork which was supplied to Southdown Motor Services in quantity between 1958 and 1967. The "Queen Mary" batch is synonymous with Southdown's operations in East and West Sussex during the 1960s and 70s, comprising different bodywork styles including convertible open-tops and buses with larger panoramic windows.

Leyland were at the forefront of introducing rear-engined double deck vehicles such as the Leyland Atlantean, with operators taking advantage of the Government's Bus Grant to introduce modern, one-man operated vehicles. The last UK example delivered was TTD386H for SELNEC at the end of 1969, although some PD3 chassis were still being exported to India into 1970.

The final PD chapter relates to the Northern General experimental one-man operation "Tynesider" bus, which was constructed in 1972 using a1949 MCW Orion bodied Leyland PD2 and the front end of an accident damaged AEC Routemaster. Re-registered to MCN30K and given Northern General fleet number 3000, the Tynesider had a full width Routemaster bonnet, and the driver's position was moved back so that it could be one man operated. Besides being an unattractive vehicle, it was not often used in service after 1972 and was eventually sold in 1978, now being privately preserved.

Bristol Lodekka (1949-1968)

Bristol Commercial Vehicles developed the Lodekka for the UK state-owned bus sector in the late 1940s, with the aim of retiring the Bristol K-type which had been in production since 1936. The name "Lodekka" was chosen to reflect the low overall height that was to be achieved using a complex drop-centre rear axle. This allowed Eastern Coach Works of Lowestoft to construct a body that included a sunken gangway on the lower deck, which in turn permitted a lower ceiling, upper deck floor and roof line, and an overall vehicle height of just 13' 5". The vast majority of Lodekkas were powered by five or six-cylinder Gardner engines, denoted by 5G or 6G at the end of the vehicle's model number, although Bristol's own six-cylinder BVW engine (6B) and Leyland's equivalent (6L) were specified by some fleets. The first two Lodekka "LDX" prototypes, LHY949 and JWT712 were delivered in 1949, built to the maximum dimensions permitted at the time of 26' long and 7' 6" wide, and providing seating for 58 passengers.

The first generation of Lodekkas, known as "LD", retained the open rear platform and entered production in 1954. These were longer than the prototypes at 27' owing to an increase in maximum vehicle dimensions, and a further relaxation in 1956 allowed Bristol to introduce a 30' "LDL" (LD Long) variant which provided an increased capacity for 70 passengers, with the existing 27' version then becoming known as "LDS" (LD Short). The second generation of Lodekkas was introduced in 1960, introducing a completely flat floor on the lower deck, and an option of open rear platform or forward entrance with doors. The new models were known as "FS" (Flat, Short) and "FSF" (Flat, Short, Forward Entrance), "FL" (Flat, Long) and "FLF" (Flat, Long, Forward Entrance).

With over 2,000 examples built, the Lodekka was used by all British Transport Commission "Tilling Group" operators and the Scottish Bus Group and was a common sight throughout the country. Bristol was prohibited from selling its Lodekka chassis on the open market, so they granted a licence to Dennis to build similar Dennis Lolines at their Guildford premises between 1958 and 1967. After almost fifteen years of production, the final Lodekkas were completed in 1968 with G-registrations, with the last examples being delivered to Eastern National and Midland General. The Lodekka gave way to the new rear-engined Bristol VR range, its modern styling ending both the half-cab format and the need to employ a conductor.

The Lodekka was used extensively in London Weekend Television's "On the Buses" comedy series between 1969 and 1973, with Eastern National's FLFs 2885 (WNO973F), 2911 (WWC741F), 2917 (AEV811F) and 2930 (AVW399F) being seen most frequently. The fictitious bus company was Luxton and District, and Eastern National's Wood Green depot was used as a filming location. A spin-off film "Holiday on the Buses" was made in 1973, which saw former Crosville Motor Services open top Lodekka LD6G XFM229 being used for holiday camp transport. Filmed at Prestatyn Pontins in North Wales, the end of the film saw the Lodekka stuck in the sand on the beach with an incoming tide, creative camera work ensuring the bus came to no permanent harm.

Above: 841SHW was new to the Bristol Omnibus Company in August 1964 as C7148. The 70 seat ECW bodywork was modified in 1976 to open top format and continued in use until withdrawal in 1982.

Lodekka numbers declined in the 1980s, many ending their days as driver training vehicles or as mobile homes. Some Lodekkas were exported to mainland Europe, as their low overall height made them more suitable for the lower European bridge clearances that were impassable by normal height double deck buses. The most widely known second owner for the Lodekka was London-based Top Deck Travel, who converted over 100 of the type to mobile "Deckerhomes" that were used on expeditions to almost every corner of the globe between 1973 and 1997. One of the most popular Top Deck expeditions was from London to Kathmandu in Nepal, a journey which took ten weeks if all went well. One of their earliest buses, PDL519 which was new to Southern Vectis in June 1958, is currently preserved in its Top Deck Travel format by the Isle of Wight Bus Museum at Newport.

Above: Eastern National Bristol Lodekka FLF6G AVX975G (2614) was new in August 1968 and was delivered as a 31' long coach with a 55 seat ECW body. Coach Lodekkas were commonly used on longer distance services and were a regular sight at London's Victoria Coach Station having operated routes from Southend-on-Sea. 2614 has recently undergone restoration back to its original condition.

Bedford SB (1950-1987)

Bedford developed the front-engined SB chassis as a replacement for the Bedford OB, which had been in production since before the Second World War. Unveiled at the Commercial Motor Show in 1950, it was the first Bedford chassis to provide a "forward control" layout, with the driving position above the front axle and alongside the engine, allowing greater use of space than had been possible with the "normal control" layout of vehicles such as the Bedford OB. This allowed for an increased passenger capacity of 33. The SB was made available with a wide variety of different six-cylinder petrol and diesel engines, including Bedford, Leyland and Perkins, coupled to a four-speed synchromesh gearbox (five speed was available as an option from 1960). With vacuum hydraulic brakes and improved suspension, the SB was a more comfortable chassis than its predecessors, and was available to be bodied as either a bus or coach by Burlingham, Duple, Harrington, Lex, Marshall, Mulliner, Plaxton, Strachans, Thurgood, Willowbrook, Wright and Yeates, although Duple's Vega and Super Vega with their "butterfly fronts" became the most popular.

The SB became a common sight in the fleet of smaller, independent bus and coach operators throughout the UK. The Ministry of Defence were a significant user of the SB for several decades, using Lex, Marshall, Mulliner, Strachans and Willowbrook bus bodied examples at defence establishments all over the country, and in some cases overseas (including Cyprus and Aden). These were typically fitted with 36 bus seats and had far less chrome fittings than privately owned coach equivalents. Several police constabularies also took examples of the SB for police officer and prisoner transport purposes. The Bedford SB was also available for the mounting of narrow 7'6" bodywork, which became popular with operators running rural services on narrow country lanes. This version became popular with coach operators on Jersey in the Channel Islands, where the maximum vehicle width is 7'6", and dozens of Duple bodied examples were used by operators such as Blue Coach Tours, Pioneer Coaches and Waverley Tours before newer narrow coaches such as the Leyland Swift became available.

The largest export customer for the Bedford SB (and in fact the largest SB fleet in the world) was operated by New Zealand Railway Board Services, who purchased 1,260 examples between 1952 and 1981, almost all of which were completed by New Zealand Motor Bodies (NZMB). Only 39 buses had Bedford's diesel engines fitted, the balance being petrol. The SB was exported to many other countries, including Africa, Australia, India and Pakistan. The SB chassis was also used for non-PSV purposes too, becoming fire engines, mobile libraries and even seven Bedford SB3 mobile cinemas for the Ministry of Technology - this last group being supplied with trailers to give demonstrations following engineering lectures in the cinema. One the mobile cinemas, KJU267E, is fortunate to have survived and is fully restored.

Above: Appleby Coaches of Louth in Lincolnshire took delivery of PFW419M in May 1974. This 41-seat coach carries Plaxton Panorama Elite bodywork on a Bedford SB5 chassis, and has been beautifully restored to its original condition.

In the 1980s, Bedford's fortunes changed as it failed to secure new contracts with the Ministry of Defence and faced competitors with more advanced vehicles. The SB was produced up until Bedford ceased commercial vehicle production in 1987.

Bedford SB Specifications

- SB/SBG/SB3 – Bedford 4.9-litre petrol engine
- SB0 – Perkins 5.5-litre R6 engine
- SB1 – Bedford 4.9-litre diesel engine
- SB5 – Bedford 5.4-litre diesel engine
- SB8 – Leyland 5.8-litre O.530 engine
- SB13 – Leyland 6.1 litre O.370 engine
- NFM – General Motors designation for petrol-engined SBs
- NJM – General Motors designation for diesel-engined SBs

AEC Regal IV "RF" (1951-1953)

The original front-engined single deck AEC Regal was first produced in 1929 and became a popular choice throughout the country. It was perhaps best known for becoming the vehicle upon which many early London Transport "Green Line" coaches were built in the period before the Second World War. The Mark IV AEC Regal, which was announced in 1948 and became available in 1950, was the first Regal to have its 9.6-litre diesel engine mounted under the saloon floor, and with over half of the London Transport single deck fleet at the time being over fifteen years old became a suitable candidate for a planned significant fleet update.

The prototype Regal IV, registered UMP227, was bodied by Park Royal Vehicles and loaned to London Transport for trials in May 1950, initially being allocated to St Albans for Country Area route 355. Based upon trials of the prototype, London Transport made the decision to order 700 chassis, which were to be bodied by Metropolitan Cammell of Birmingham, a decision made considering that both Park Royal and Weymann were at the time operating at capacity producing bodywork for their AEC Regent III "RT" buses. The Regal IV was designated the "RF" class, and the vehicles were all delivered between 1951 and 1953. The first twenty-five of the batch (RF1-25) were needed urgently for private hire work and the 1951 Festival of Britain in particular. They were 27'6" long (the maximum length permitted at the time) and were fitted with roof mounted curved glazing to increase the passengers' visibility. The next batch of 263 Regals (RF26-288) consisted of 30' long 39-seat Green Line coaches which started to be delivered later in the same year. A fleet of 30' red liveried Central Area RF buses followed next (originally RF289-513), being specified without doors at the request of the Metropolitan Police to hasten passenger loading times and hence reduce traffic congestion at bus stops. The final RF delivery of 187 was of green 30' buses for Country Area services (originally numbered RF514-700). During the life of the RF, there were numerous examples of modifications, re-paints and re-numberings, to meet changing route requirements and cascading caused by the subsequent arrival of newer vehicles.

Based upon the standard RF design, London Transport convinced British European Airways to order 65 AEC Regal coaches in 1952/53 with a distinctive Park Royal "deck and a half" bodywork. These were used by London Transport on BEA contracts as passenger transfer buses between their Central London terminal and London's airports at Heathrow, Croydon and Northolt. The vast under floor space of these high floor coaches was used for passenger luggage. As aeroplanes became larger in the late 1950s and early 1960s, these 18-seat coaches lacked the capacity to meet growing passenger numbers, which lead to the decision to purchase front-entrance "RMA" Routemasters, many of which operated towing luggage trailers. The BEA RFs were withdrawn in 1966/67, many being scrapped shortly afterwards. Three were bought for use as international touring coaches by Continental Pioneer of Richmond (MLL 719/720/747, with NLP643/8 acquired for spares), and four were retained by London Transport (MLL725/7/9,735) and modified to become mobile staff uniform issue units, each of which towed a trailer until their withdrawal in 1977/78.

Withdrawals of RFs commenced in the early 1960s as new AEC Routemaster coaches became available. The formation of the National Bus Company created London Country Bus Services in 1970, which inherited all "green" RFs including the Green Line coaches and Country Area buses. Their numbers started to decline as London Country's large orders of Leyland Nationals were delivered and they were finally withdrawn from bus service in May 1978 (Chelsham's RF684), although Northfleet's modernised Green Line coach RF202 continued in use until May 1979, when it too was withdrawn and officially preserved alongside the company's Routemaster coach RMC4. The number of red Central Area RFs also declined steadily as new Leyland Nationals were introduced, but the small engineering pit at Kingston Garage could only accommodate the RF so the survivors gravitated to this location. Central Area RFs were ceremoniously withdrawn in March 1979, with Kingston's RF507 being the last to be used in passenger service. London Country continued to operate a trio of RFs (79, 556 and 647) as tow trucks until the early 1980s, and RF594 was used as a dedicated recruitment bus from 1973 until 1981.

Above: RF13 was one of 25 "private hire" AEC Regal IV coaches delivered to London Transport in 1951 which were initially used for the 1951 Festival of Britain events. It was withdrawn from service in 1963 and joined the fleet of Hampsons of Oswestry in 1964, in whose livery it has since been preserved.

Surprisingly, and possibly due to their long period of use by London Transport and London Country, many withdrawn RFs were promptly scrapped by Wombwell Diesels or Booths of Rotherham, and only a small number saw further PSV use with smaller independent operators such as Continental Pioneer, Hampsons of Oswestry, Orpington & District, Osbornes of Tollesbury, Premier Travel of Cambridge and Silverline of Hounslow. BEA at Heathrow Airport purchased several for use as airside buses, for which role each received a two-way radio and a red warning light on the roof, and others were used by Scout Groups, schools, gliding clubs and as mobile homes.

Above: London Transport's Green Line coaches received 263 AEC Regal IVs between 1951 and 1953. RF269 started work from Amersham Garage in June 1952, being downgraded to bus specification in 1965. It joined the newly-formed London Country bus fleet in 1970, and was finally withdrawn and sold in 1972.

One notable RF survivor was RF255 (MLL792) which had passed through a number of private owners since being withdrawn by London Country in January 1976. By March 1994, extensive restoration work had been completed and the bus once again gained a PSV certificate, and it then ran the Metrobus Sunday "Wealdsman" service 746 between Bromley and Tunbridge Wells for the 1994 and 1995 seasons. For this duty, the RF received a smart deep blue livery with yellow relief, in line with Metrobus corporate colours, and carried yellow side route boards. RF255 remains preserved and joins the ranks of dozens of privately restored RFs that can be seen at events and rallies and in museums up and down the country.

AEC Reliance (1953-1979)

AEC of Southall in London developed the Reliance chassis in the early 1950s, incorporating a mid-underfloor mounted engine also built by AEC. Whilst the initial engine was a 7.7-litre AH470, subsequent changes saw this rise through the 8.1 litre AH505, 9.6-litre AH590 and 11.3-litre AH690/1 to the largest power unit, the 12.4-litre AH760. Synchromesh gearboxes were offered by AEC and ZF, with AEC also offering an epicyclic semi-automatic transmission. Two prototypes were completed in 1953: 135BMV was fitted with a centre entrance Duple 41-seat coach body, whilst 50AMC was finished as a forward entrance 44-seat bus by Park Royal.

Whilst the Reliance was originally considered a lightweight chassis at 30'0" length, changes to the Construction and Use Regulations in 1961 allowed for single deck vehicles to be built to a maximum length of 36'0" and AEC soon took advantage of this with a longer, heavyweight chassis powered by the AH590 engine. This coincided with the opening of the UK motorway network, and express service operators were amongst the first to realise the value of this new version. AEC was acquired by Leyland Motors Limited in 1962.

The variety of different bodywork available for the Reliance ensured that there was an option suitable for every operator's individual needs, with offerings from Alexander, Beadle, Bellhouse Hartwell, Crossley, Duple, East Lancs, Harrington, Mann Egerton, MCW, Plaxton, Roe, Strachan,

Weymann, Whitson, Willowbrook, Windover and Yeates, amongst others. Wallace Arnold of Leeds was an early adopter and is remembered by many for their characteristic Burlingham bodied coaches with central entrance door. Other significant Reliance fleets were operated by Aldershot & District, Devon General, East Kent, Northern General, North Western, Potteries, Scottish Omnibuses and Western Welsh. Export destinations included Australia, New Zealand and Trinidad and Tobago.

Above: 200APB is a Burlingham-bodied AEC Reliance delivered new to Safeguard of Guildford in 1956. After withdrawal in November 1962, it was purchased by Safeway Coaches of South Petherton in Somerset. 200APB was eventually repurchased by Safeguard in 2003 for preservation and occasional private hire.

In 1965, London Transport started to update their ageing Green Line coaching fleet with Reliances, using a Willowbrook bus style body with coach seating. They were, unfortunately, mechanically unreliable and spent much of their short career in store. In 1971, shortly after London Country Bus Services took ownership of Green Line, a fleet of 90 Park Royal bodied Reliances arrived, their more powerful and reliable AH690 engines provided nearly ten years' service. Their replacements were yet more Reliances, a batch of 150 with the largest AH760 engine and bodied by either Duple at Blackpool or Plaxton at Scarborough. These were the first recognisable coach bodies to be used on Green Line services, and created a modern image later continued by Leyland Tigers.

Above: London Country purchased five AEC Reliance 6U3ZRs with Plaxton Panorama Elite coach bodywork in 1973, which were used on National Express, National Holidays and private hire work. P3 worked from many London Country garages, and has since passed into preservation carrying this 1977-style of Green Line livery.

The Reliance was the last public service vehicle to be produced by AEC. Reliance production ended in early 1979, and after final deliveries in 1980, many operators switched to Leyland Leopards or Volvo B58s for their future coach needs. The very last Reliance bus was Duple Dominant bodied JTM109V, which was delivered new to Tillingbourne, and passed to Metrobus when Tillingbourne's Surrey and Orpington operations were split. It subsequently passed to Sussex Bus and then Guildford & West Surrey before being restored in Metrobus livery by a group of private preservationists.

The Reliance was the last public service vehicle to be produced by AEC. Reliance production ended in early 1979, and after final deliveries in 1980, many operators switched to Leyland Leopards or Volvo B58s for their future coach needs. The very last Reliance bus was Duple Dominant bodied JTM109V, which was delivered new to Tillingbourne, and passed to Metrobus when Tillingbourne's Surrey and Orpington operations were split. It subsequently passed to Sussex Bus and then Guildford & West Surrey before being restored in Metrobus livery by a group of private preservationists.

Guy Special "GS" (1953-1956)

At the beginning of the 1950s, London Transport was looking for a small bus which could be used to replace its 1935/36 Leyland Cubs operating on rural bus services which were ideally suited to using narrow and winding country lanes. The new AEC Regal IV "RF" class buses were too large to be suitable for this purpose. The outgoing Cubs were one person operated, as their routes would have become uneconomical if they had to be crewed by two staff, so the replacement vehicles would need to be operated in the same way.

London Transport specified that the new vehicle was to be produced on a standard chassis, the Guy Vixen (constructed by Guy Motors of Wolverhampton in the West Midlands), which was powered by the Perkins P6 4.73-litre diesel engine, producing 65 bhp, and being connected to a four-speed constant mesh crash gearbox. The distinctive bodywork was constructed by Eastern Coach Works of Lowestoft, incorporating a protruding bonnet, enclosed front wings and being finished off with Guy's Indian figurehead. The design intentionally incorporated many body and electrical components from the RF class, helping in their maintenance and repair. The licensing authorities allowed the Guy's seating to be increased to 26 (from the Leyland Cub's 20) whilst still permitting one-man operation, this capacity being achieved by the fitting of bench seats above the rear wheel arches.

Above: Guy Vixen Special GS2 was new to London Transport in October 1953, with ECW 26-seat bodywork. It was initially allocated to Hitchin Garage for route 383 between Hitchin and Weston and moved to Stevenage Garage in 1959. It was withdrawn and sold in 1962

In 1952, Guy manufactured eighty-four Vixen chassis, which received their ECW bodywork and started to enter service in October 1953. Receiving fleet numbers GS1-GS84 (and corresponding registrations MXX301-MXX384), GS84 was the last to enter service in January 1956. Operation of the type was concentrated on approximately 20 of London

Transport's Country Area garages, in particular those that operated rural routes which were not suitable for the allocation of normal sized buses, although some garages only operated the type for short periods. The Guy Specials enjoyed a comparatively short operating life before they started to be withdrawn. The 1962 batch of Country Area Routemasters cascaded AEC Regal IV "RF" buses to rural services, and October marked the start of their demise with Amersham, Chelsham and Epping garages losing all their allocations, with further withdrawals continuing over the next few years. When London Transport's Country Area became part of London Country Bus Services in 1970, all remaining Guys were transferred to the new operator apart from five which were retained by London Transport for staff transport duties to their Aldenham and Chiswick Works.

London Country's last GS operation was on 29th March 1972, using GS33 and GS42 on route 336A between Rickmansworth and Loudwater. After withdrawal, the type was eagerly acquired by several independent operators, most notably Tillingbourne Valley in Surrey who took thirteen and Southern Motorways in West Sussex who purchased seven (and one for spares). GS40 and GS41 were sold to West Bromwich Corporation in 1961, GS24 was bought by the London Fire Brigade (before later also passing to Tillingbourne Valley), and GS47 joined St John Ambulance in London. Several became mobile caravans, the most unusual being GS50 (Q75KUA), which had its ECW body mounted onto a Ford Cargo chassis in the 1990s after the original Guy chassis had rotted away, and subsequently made return trips to Romania and Morocco. Currently, at least ten GS types are actively preserved and a further three of the type are thought still to exist in private ownership in Belgium.

AEC Routemaster (1954-1968)

The world famous Routemaster was developed jointly by AEC of Southall and London Transport. Design work commenced shortly after the end of the Second World War in 1947 and the first prototype, RM1 (SLT56), was displayed at the Commercial Motor Show in September 1954. The new vehicle had to meet many key design criteria, including being lighter and hence more fuel efficient than the existing AEC Regent

III "RT" buses, being easier to operate, and capable of being maintained at London Transport's new Aldenham Works at Elstree in Hertfordshire, which opened in 1956. They were also required to be a replacement for the retiring London trolleybus fleet, the last of which were withdrawn in May 1962. The Routemaster was designed as an integral bus, with the front frame (carrying the engine, steering and front suspension) being connected by the bodywork to the rear frame (carrying the rear axle and rear suspension). The design used lightweight aluminium, and was revolutionary in incorporating power steering, powered braking and an automatic gearbox. London Transport specified four different Routemaster prototypes:

- **RM1** (SLT56), AEC AV600 9.6-litre, Park Royal bodywork (entered passenger service in February 1956)
- **RM2** (SLT57), AEC AV470 7.7-litre, Park Royal bodywork (used for trials until passenger service in May 1957)
- **RML3** (SLT58), Leyland O.600 9.8-litre, Weymann bodywork (entered passenger service in January 1958)
- **CRL4** (SLT59), Leyland O.600 9.8-litre, ECW coach bodywork (entered passenger service in October 1957)

RM8 (VLT8) was also used as an engineering test and development vehicle and did not enter passenger service until 1976. Full Routemaster production commenced early in 1959, with existing test rigs RM5-7 (VLT5-7) finally receiving Park Royal bodywork and entering passenger service. Standard Routemasters were 27'6" long and powered by an AEC 9.6 litre engine. They were fitted with 64-seat bodywork with an open rear platform provided by AEC subsidiary Park Royal Vehicles, seating 28 on the lower deck and 36 upstairs. The longer "RML" at 30' seated 72. The standard-length coach version for Green Line services was the "RMC", seating 57, and the longer 30' coach version was the "RCL" which seated 69. Both the RMC and RCL coaches were fitted with electrically operated rear platform doors. A unique forward entrance Routemaster (RMF1254, 254CLT) was built in 1962, and whilst it never turned a wheel in London Transport service, it did pave the way for the development of the forward entrance Routemaster, resulting in two sizeable orders.

Above: SLT59 was one of the four Routemaster prototypes and was originally allocated fleet number CRL4 ("Coach Routemaster Leyland") when delivered in June 1957. It was fitted with a 57-seat coach body by ECW, which included rear platform doors. It was extensively trialed in the early 1960s, becoming RMC4 in 1961, and later passing to the newly formed NBC-subsidiary London Country Bus Services in 1970.

In 1964/65, Northern General ordered 50 "RMFs" for fast inter-urban services, for which they were fitted with Leyland engines and a Monocontrol semi-automatic gearbox, many carrying the "Shop at Binns" logo above their front destination box. One of the batch (RCN685) was withdrawn after a serious accident in 1972, and was subsequently rebuilt with the drivers cab and staircase moved further back, allowing for experimental one man operation. Northern General named this unusual vehicle the "Wearsider", which continued to be used until 1978 when it was withdrawn and scrapped.

Above: KGJ601D was one of 65 front-entrance Routemasters delivered to British European Airways in 1966/67. Many towed luggage trailers, "BEA1" clearly showing the towing bracket. This example received BEA's orange and white livery in 1970 and then British Airways colours in 1974, before being sold to LT in 1979.

British European Airways (BEA) ordered 65 "RMAs" in 1966/67 for the operation of passenger transfer services between their Central London Terminal in Cromwell Road and Heathrow Airport. These were fitted with upgraded transmissions to allow a lively 70mph performance on motorways, and many operated with one of the 88 specially constructed Marshall luggage trailers attached to the rear. The RMAs started to be withdrawn in 1975 and were subsequently purchased by London Transport. They proved not to be successful in passenger service, so ultimately were retired to driver training duties or were used for staff transport to the Aldenham and Chiswick Works, a task they continued to perform until December 1987. One additional forward entrance

Routemaster worthy of note is FRM1 (KGY4D), which was a 1966 prototype with front entrance and rear engine, constructed using 80% standard Routemaster components and an AEC AV691 engine. Although popular, it was not put into production, but fortunately was saved and is now part of the London Transport Museum collection.

The modular construction of the Routemaster allowed for efficient overhauls to be carried out at the Aldenham Works. Upon arrival, the bodywork and chassis were separated and sent to different sections, with any remedial work needed to the chassis and operational components (engine, gearbox etc) being undertaken at London Transport's Chiswick Works. The bodywork remained at Aldenham and was initially inverted to allow a full steam clean of the underside before being sent for body panel repair or replacement, and the repair or replacement of any damaged interior fitments. Due to the different lengths of time needed for each stage, and the chassis being overhauled in a shorter time than the bodywork, Routemasters were rebuilt using the next available chassis, engine, gearbox and bodywork combination that became available, and only a very few Routemasters were ever reunited with their original Park Royal bodywork. The completed bus would then be test driven on roads around the Aldenham site, and if it passed mechanical checks it was then sent for a repaint, the fitting of re-covered seats, before finally being recertified and delivered back to a London Transport garage for its next period of bus operation.

The Routemaster fleet was split in 1970 with the formation of London Country Bus Services, with the green Country Area buses and Green Line coaches passing to the new fleet. After only a few years, their RCL and RMC coaches were the first to be withdrawn as orders for newer vehicles were delivered, the earliest being RCLs in 1975 at just ten years old, although some were subsequently repurchased by London Transport. It was the RM and RML buses that became the mainstay of the Central London fleet in the deregulation bus service era, and some Routemasters found themselves changing owners as other operators started to win tendered London bus contracts. Kentish Bus took over operation of Route 19 in April 1993 and leased 24 refurbished RMLs which had been fitted with Iveco engines and painted into Kentish Bus maroon and cream colours. In 1997, Route 19 passed within the Cowie

Group to South London, and these RMLs were painted back into London red. 1993 also witnessed BTS of Borehamwood securing the operation of Route 13, which resulted in the leasing of 22 RMLs from London Buses for its operation. In 1994, the component London Buses companies were privatised, and the Routemaster fleet was split accordingly amongst the nine new owners.

Above: WLT577 is a 1961 AEC Routemaster. Following an overhaul at Aldenham Works in 1983, RM577 was matched with the bodywork from RM664. The latter was unusual in being an experimental unpainted Routemaster to assess weight and cost saving and was affectionately known as "The Silver Lady". The trial was not successful, and RM664 was repainted red. It was withdrawn in 1987.

Always a popular and reliable bus, the Routemaster outlasted more modern vehicles that had been subsequently introduced. At the turn of the century, its numbers declined as more London routes were converted to modern, one man operated and accessible low floor buses. Withdrawn Routemasters were eagerly purchased by new owners in the post 1986 deregulated bus industry, which saw them continue to provide a useful service well beyond their original London homeland. Amongst the first to purchase the Routemaster was Perth based Stagecoach Holdings, which used them on low cost, competitive "Magic Bus" services in the Glasgow area. Southend Transport purchased 23 Routemasters (plus three for spares) in 1991 and operated these until 1993 with many then passing on to another new Routemaster operator, Reading Mainline. Established to compete with Reading Buses, Reading Mainline operated an all Routemaster fleet from July 1994 until eventually being bought out by the larger operator in 1998. A condition of the sale was that the Reading Mainline identity was to remain for two years, so Routemaster operations in Reading lasted until July 2000, at which point many were repurchased by Transport for London (see below). Other notable Routemaster operators included United Counties, who acquired eight for service 101 in Bedford and East Yorkshire Motor Services who operated Routemasters on bus services in the Hull area. Scottish Routemaster fleets included Clydeside Scottish and Kelvin Scottish, who amassed considerable Routemaster fleets to compete with Strathclyde Buses in Glasgow, and Strathtay Scottish who used Routemasters in Perth and Dundee, the former against heavy competition from Stagecoach who were coincidentally using similar Routemaster buses. Perhaps the most distant Routemaster acquisition was by the Sri Lankan Central Transport Board, who purchased 40 refurbished Routemasters in December 1988.

A "second chance" was granted to former London Routemasters in 2000, when newly elected London Mayor Ken Livingstone set about reacquiring examples from fleets including Halifax JOC and Reading Mainline, as well as a number which had been owned by private preservationists. Starting in 2001, these "new" arrivals were sent to Marshalls of Cambridge for modernisation, which included the fitting of Cummins engines, Allison gearboxes and Voith retarders, new interior and exterior lighting, new body panels and repainting, new interior seat coverings and yellow grip applied to all grab rails. When completed, they were allocated to Arriva London (for route 38), London Central (for route 36) and Sovereign (for route 13). Despite these overhauls, routine operations of the London Routemaster ended in December 2005 after nearly fifty years, with Arriva London South's RM2217 operating the last 159 service on 9th December 2005. A small number of modernised Routemasters continue to work in London on "heritage routes" 9 (Kensington High Street to Trafalgar Square, operated by First London) and 15 (Tower Hill to Trafalgar Square, operated by Stagecoach London), both services being short workings of the full Transport for London services.

Of 2,876 Routemasters built, over 1,000 Routemasters still exist to this day, with many hundreds in the hands of preservationists. Other examples can be seen in countries all over the world, as sightseeing buses, publicity vehicles or museum exhibits. The iconic Routemaster retains its popularity, and its styling was a major influence in the design of Transport for London's "New Bus for London", which was built by Wrightbus and which entered passenger service in London in February 2012.

AEC Routemaster Specifications

- RM – standard bus (27'6")
- RML – long bus (30'0")
- RMC – standard coach (27'6")
- RCL – long coach (30'0")
- RMF – front-entrance (prototype RMF1254 and Northern General buses)
- RMA – front entrance (British European Airways buses)
- ERM – extended five-bay bus body (10 for London Coaches)
- FRM – FRM1 (KGY4D), front entrance, rear-engined prototype

Albion Nimbus (1955-1965)

Leyland Motors had acquired the Glasgow based manufacturer Albion Motors in 1951 and focussed their attentions on the production of lightweight bus and coach chassis for UK operators. Albion developed two lightweight engines for this task, the EN219 and EN250, and the EN219 was introduced into Albion's Claymore delivery truck in 1953. Scottish Omnibuses developed a 32-seat bus in 1954 based upon the

Claymore truck chassis, and the resulting Albion Nimbus was unveiled at the Scottish Motor Show in 1955.

Above: Nimbus MR9N NSG869 was built for Albion Motors as a prototype in 1955 and was fitted with an 8' wide Scottish Omnibuses style 32-seat front-entrance body.

The original Nimbus was the MR9N which was powered by the 3.8-litre EN219 engine driving a four-speed synchromesh Albion gearbox, and in 1958 the Nimbus NS3N was fitted with the larger 4.1-litre EN250 engine and a David Brown four-speed constant mesh gearbox.
The final version launched in 1960 was the Nimbus NS3AN, which now offered an Albion five or six speed constant mesh gearbox and vacuum assisted braking.

Only really being suitable for lightly used services in rural areas, the Nimbus was not purchased in great numbers, although it did enter the fleets of BET and Scottish Bus Group companies, municipal operators and independents alike. The largest Nimbus fleets were operated by Western Welsh (48 Harrington and Weymann bodied), Walter Alexander & Sons (15 Alexander bodied), Maidstone & District (15 Harrington bodied), Halifax Corporation (10 Weymann bodied) and Devon General (9 Willowbrook bodied). The narrow width and short wheelbase of the Nimbus made it ideal for services on the narrow and winding roads of the Channel Islands, and the combined Guernsey Motors and Guernsey Railways bus fleets ordered 32 with Reading bodywork. Smaller independent operators commonly specified Plaxton bodywork for Nimbus coaches. International orders saw the Albion Nimbus being shipped as far afield as Argentina, Australia, Denmark, Netherlands, Portugal, South Africa and Sri Lanka.

Albion stopped producing the Nimbus in 1965, having delivered 124 of the MR9N version and 217 NS3N/NS3AN types.

Leyland Royal Tiger Worldmaster (1956-1964)

As a successor to the underfloor-engined Leyland Royal Tiger, the Royal Tiger Worldmaster (more commonly known simply as the "Worldmaster") was powered by a Leyland O680H engine connected to a pneumocyclic semi-automatic gearbox. Production commenced in 1954 and continued for 25 years until 1979, by which time more than 20,000 had been constructed – making it Leyland's most successful bus chassis. However, almost all Worldmasters were sold abroad with only a very small number of these heavyweight chassis finding their way into UK fleets – most operators preferring the Leyland Leopard L1/L2 which was introduced in 1959.

Glasgow Corporation took 30 (LS1-30) in 1956 (to be fitted with Weymann B40D body frames) and Halifax Corporation a further ten (KCP1-9) in 1958 – with completed Weymann 42-seat bus bodies - these 39 being the only bus-bodied Worldmasters used in the UK. All other domestic Worldmasters were bodied as coaches, including for Ellen Smith Coaches of Rochdale (Plaxton Consort C41C-bodied ODK137 and SDK442) and Gliderways of Smethwick (Harrington C37C-bodied YHA26-28). It has been recorded that a further chassis destined for export was purchased in 1963 by Happiways of

Manchester, receiving a Duple Northern body. At the end of the 1960s, Ellen Smith Coaches rebodied two of their Worldmasters to Plaxton Panorama (ODK137) and Plaxton Elite (SDK442) C41F-layout coaches.

Above: SDK442 is a Leyland Royal Tiger Worldmaster new to Ellen Smith Coaches, which was later rebodied with a Plaxton Elite C41F body. It is preserved in Greater Manchester's Museum of Transport.

Internationally, the Worldmaster could be seen hard at work almost everywhere. Over 5,000 saw service in Israel, with 70% of these built locally by Leyland and in Spain and Portugal some were fitted with double-deck bus bodies. State-operator Rhodesian Railways specified a three-axle configuration for a batch of their Worldmasters. Nearer to home, Córas Iompair Éireann ordered 17 to be fitted with CIE coach bodywork in 1962/3 – commencing with WT1 (HZD580). Like in the UK, the resilience of the Worldmaster chassis made them ideal for rebodying in their later years and Egged in Israel upgraded 40 of their early buses in the 1980s. CIE also rebodied its Worldmasters with new VanHool Vistadome bodies in 1970, becoming the WVH class in the process.

The Worldmaster was discontinued in 1979, with the existing Leyland Leopard replacing it.

Leyland Royal Tiger Worldmaster Specifications

- RT1 - 35' (10.67m) long, 20' (6.10m) wheelbase
- RT2 - 33' (10.06m) long, 18' 6" 5.64m) wheelbase
- RT3 - 30' (9.14m) long, 16' 2" (4.93m) wheelbase
- E prefix – chassis built for export
- C prefix – chassis built with low ground clearance
- L prefix – left-hand drive (e.g. LERT1, LCRT2)

Leyland Atlantean (1956-1986)

Following the Second World War, there was an interest in developing larger capacity vehicles where the engine could be moved away from alongside the driver. Several under-floor designs were explored, although this invariably raised the floor height resulting in more steps for the passengers to climb. In 1952, Leyland produced a rear-engined prototype, powered by a Leyland O.530 engine and bodied by "SARO" (Saunders-Roe) with a 7'6" wide design, registered STF90 and which carried the designation "PDR1" (PD "Rear"). A second prototype followed four years later: XTX684 was powered by an identical engine, and its Metro-Cammell bodywork seated 78 passengers. This second prototype was known as the Leyland Lowloader. Unfortunately, both prototypes had rear passenger entrances which wasted the space alongside the driver.

1956 saw a change in the Construction and Use Regulations, allowing double deck vehicles to be 30' (9.1m) long. This allowed for greater space ahead of the front axle, and Leyland took advantage of this to develop the front-entrance Leyland Atlantean. The first prototype, 281ATC, was displayed at the 1956 Commercial Motor Show in London, and whilst similar to the Lowloader it used a strong and light fabricated frame, light alloy floor plates and a drop-centre rear axle allowing a completely flat floor. One initial challenge was unacceptable engine noise in the lower saloon, as the rear-mounted engine had originally been designed to sit within the bodywork shell.

The production of the Atlantean "PDR1/1" commenced in 1958. The engine was now outside of the bodywork in its own compartment, and the low floor was initially sacrificed to allow the use of conventional axles. The drop-centre rear axle became available once again in the 1964 "PDR1/2" model, and a further increase in maximum vehicle length in 1967 produced the 33' (10m) long Atlantean "PDR2/1". A new design of Atlantean was unveiled in 1972: the "AN68" with Leyland's new O.680 engine was offered in two lengths, the AN68/1R at 9.4m and the AN68/2R at 10.2m. The new model offered a wider entrance and improved mechanical systems. In 1978, the "AN69" model was added powered by the new Leyland O.690 turbo-charged engine, all but one of these being supplied to overseas fleets in Teheran (Iran), Pretoria (South Africa) and Quito (Ecuador).

Whilst the Bristol VR became the default double deck vehicle for many National Bus Company subsidiary companies, the Atlantean found popularity with municipal fleets including Glasgow, Edinburgh, Liverpool, Manchester, Nottingham and Plymouth. Some NBC fleets chose the Atlantean over the Bristol VR – London Country was one of them, amassing a fleet of nearly 300, whilst others such as Southdown operated a mixture of Atlanteans and VRs side by side. The Atlantean carried many different bodies, including Alexander, Duple Metsec, ECW, Northern Counties, Park Royal and Roe, all having a distinctive external rear recess on the lower deck to allow for the lift-up engine cover.

Above: Kingston upon Hull Corporation Transport Leyland Atlantean 270 was one of twenty delivered in December 1969 and January 1970. It was supplied with dual-door bodywork built by Roe of Leeds, seating 43 passengers on the upper deck and 28 on the lower.

One of the first Atlantean users was Ribble subsidiary Standerwick, who developed their "Gay Hostess" motorway coach brand in 1959 ahead of the service launch in April 1960. Ribble 1251 (MCK812) became the prototype, seating 50 passengers in comfort (34 on the upper deck, 16 on the lower) within a Weymann body constructed on MCCW framework and including a passenger toilet. 10 similar Atlanteans were delivered in 1960 and a further 12 in 1961, and these continued to operate London express services using the newly opened M6 and M1 motorways from the North West until 1968, when they were replaced by Bristol VRs.

Above: Greater Glasgow PTE Leyland Atlantean HGD903L was new in July 1973 to Newlands Garage, and was fitted with Alexander dual-door bodywork seating 45 passengers on the upper deck and 31 on the lower. LA697 was withdrawn in September 1980 and was then used as a driver trainer. It was acquired for preservation in 1998.

Unusually, a few Atlantean chassis were originally bodied as single deck buses, rather than being subsequently converted as a result of low bridge incidents. The most notable orders were three Marshall bodied buses for Great Yarmouth Corporation in 1968, and ten Pennine Coachcraft bodied buses for Portsmouth Corporation in 1972, these being even more unusual in being allowed to tow luggage trailers when working transfer services from The Hard Interchange to the Sealink ferry terminal. Much later on, East Lancs offered their single deck "Sprint" bodywork as a bodywork upgrade for ageing Atlanteans, this refurbishment attracting small orders from Southampton City Transport in 1991 and Nottingham City Transport in 1993, as well as several independent operators. Fylde Borough Transport had four 1971 former Bradford City Transport Atlanteans rebodied with Northern Counties Paladin bodywork as 42-seat single deck buses in 1993, which subsequently joined the Blackpool fleet when Fylde was acquired in 1994.

Although Leyland Olympian production had commenced in 1980, the Atlantean continued to be produced alongside its replacement until 1986, the last UK model being B75URN completed in October 1984 for Fylde Borough Transport. After that, the last deliveries of Atlanteans were exported to Kuwait Transport Company, PPD in Jakarta (Indonesia) and Singapore Bus Services. The Leyland Atlantean AN68/2R had been a surprising success in the Far East with Singapore Bus Services. An initial batch of 20 (twelve with Metal Section bodies, eight bodied by BACo) were introduced in June 1977, and were considered such a success that an order for a further 200 chassis was placed in 1978. These were fitted with bodywork at SBS premises, half being completed with Duple Metsec bodies and the rest Alexander L-type, and all entered service in 1980. Another 100 Alexander L-type bodied buses entered service in 1982. With the SBS Atlantean fleet already at 320, another 200 were ordered from Leyland, and even though UK demand for the Atlantean was waning, these were completed, fitted with Alexander R-type bodywork, and entered service between 1984 and 1986. These were the very last Leyland Atlanteans to be built.

Leyland Atlantean Specifications

- PDR1 – PD Rear Engine (first prototype)
- PDR1/1 – 30' length, conventional front and rear axles
- PDR1/2, PDR1/3 – 30' length, drop centre rear axle
- PDR2/1 – 33' length
- AN68/1R – 9.4 metre length, Leyland 0.680 engine
- AN68/2R – 10.2 metre length, Leyland 0.680 engine

Bedford J2 (1958-1976)

Bedford's J2 passenger coach was constructed using a Vauxhall 2-ton truck chassis with a 9' 11" wheelbase, modified to a "forward control" format with the driver sitting alongside the engine, instead of the "normal control" format of the original bonneted Bedford truck. The J2's power was provided by either a 3.5-litre, six-cylinder Bedford petrol engine (designated J2SZ2) or a 3.1-litre four-cylinder Bedford diesel engine (designated J2SZ10). Unusual for public service vehicles of the time was the use of 12-volt electric components and the fitting of vacuum hydraulic brakes.

Above: MVE400H is a Plaxton Embassy bodied Bedford J2, delivered new to the Cambridgeshire Regional Health Authority in 1969. It was used at the Tower Hospital in Ely, and had its seating reduced to

eleven in order to incorporate a wheelchair lift for disabled passengers. It is currently preserved by Memory Lane Coaches.

By far the most common bodywork fitted to the Bedford J2 was Plaxton's Embassy, a scaled down version of the bodywork being fitted to full-size Bedford chassis in the early 1960s. Those built up to 1965 were only 7' 6" wide and could be identified by a two-piece front windscreen and curved rear windows. Models built after 1965 were 8' wide and benefited from an inter-changeable one-piece front and rear screen. The wider bodies had a clearly visible overhang beyond the narrow chassis. 20 coach seats were fitted, comprising seven pairs, five at the rear and one forward of the entrance door next to the driver.

Whilst the Plaxton Embassy bodywork was favoured by many coach operators for the transportation of small groups, Duple Midland and Willowbrook each produced a more basic bodywork style which was mainly targeted at various UK government fleets, including the Ministry of Defence for personnel transport for the Royal Air Force and Royal Navy. Duple Midland bodies featured a single outward opening entrance door, whilst Willowbrook provided a two-piece folding format suitable for manual operation by the driver using mechanical linkage. Both versions were fitted with plastic covered bus seating, although cloth covered coach seats were available as an option. One additional bodywork variation was offered by Caetano of Portugal in the mid-1970s, who supplied their "Faro" coach bodywork to operators via the Moseley dealership. This was a heavy body, which consequently caused poor engine performance from the Bedford diesel engine and poor braking and meant that they were only ordered in small numbers

Notable operators of the Plaxton-bodied Bedford J2 included Rickards of London, who operated fifteen on airport hotel shuttles to London's Heathrow Airport for a short period in 1969/70 and were almost certainly the largest single fleet of the type. The Cambridgeshire Regional Health Authority operated four as non-PSVs, including three for the Blood Transfusion Service. A single Duple Midland bodied example (new as KLP1D) was built to a high specification for transporting staff of the Royal Household in London. The Armed Forces built up sizeable fleets of both Duple Midland and Willowbrook bodied examples, which were allocated to military installations both within the UK and overseas.

Today, the remaining Bedford J2s are a popular choice for preservation due to their small size and relatively straightforward mechanics, and there are still a small number in use as caravans by travellers.

Dennis Loline (1958-1966)

The popular Bristol Commercial Vehicles "Lodekka" was only available to be sold to the British Transport Commission's "Tilling Group" operators and the Scottish Bus Group, and so Dennis was granted a licence to produce a similar vehicle for other municipal operators seeking a low-floor double deck vehicle, and this was named the Dennis Loline.

Above: Loline III 488KOT was new to Aldershot & District in 1964, having a Weymann forward entrance body with seating for 68

passengers and being powered by a Gardner 6LW engine. Since preserved, it is seen on its home ground near Farnham, Surrey.
The chassis was manufactured at the Dennis Brothers factory in Guildford, Surrey and was fitted with a Gardner 6LW or 6LX engine normally paired with a Dennis four or five-speed gearbox. Some later Dennis Lolines were fitted with Bristol five speed gearboxes.

The Dennis Loline was produced as three different versions during its short production life, all of which were designed for two-man crew operation. Introduced in 1958, the majority of Dennis Loline I buses were delivered to Aldershot & District, who specified East Lancashire wooden framed bodywork seating 68 passengers and electrically operated rear platform doors (they took delivery of the first production Loline, SOU444). Other orders were placed by Leigh Corporation, with East Lancashire bodywork for 72 passengers and Lancashire United with Northern Counties bodywork for 69 passengers, both batches having open platforms.

Introduced in 1958, but not ordered in great numbers until 1960 was the Dennis Loline II, which brought the passenger entrance to the front and allowed the driver a clearer view of passengers boarding and alighting. The most significant orders for this version were from Walsall Corporation (Willowbrook bodied, seating 74), North Western (East Lancashire bodied, seating 71), and Middlesbrough Corporation (Northern Counties bodied, seating 70). The 1961 Loline III was the most popular and fuel efficient of the Loline models, securing repeat custom from Aldershot & District and North Western, although this time using mainly metal framed Alexander bodywork due to production issues at East Lancashire. Other corporations ordering the Loline III included Reading (East Lancashire), Halifax (Northern Counties) and Luton (with unusual Neepsend bodywork). Dennis registered a single Loline III demonstrator in 1964, using a cancelled export order vehicle built in 1962 with a 76-seat Northern Counties body: EPG179B was initially trialled by Halifax and Southampton Corporations before being sold to Warners of Tewkesbury in 1966.

Only two Lolines were sent overseas. Northern Counties bodied Loline III 8760AZ became Belfast Corporation 552 and was used for evaluation purposes from May 1961 until being scrapped in the early

1970s. Similar Northern Counties bodied Loline III AD4527 became "LW1" in the fleet of China Motor Bus in Hong Kong in January 1963, being the first double deck vehicle to work on Hong Kong Island, and was finally withdrawn from service in 1978. Of the 280 Dennis Lolines built in Guildford, 141 were purchased locally by Aldershot & District: many passed to Alder Valley upon the formation of the National Bus Company and continued in use until 1980 when they retired to make way for one-man-operated double deck vehicles. Thankfully seven of these Dennis Lolines are preserved by two local organisations and are regular visitors to rallies and events in the South of England.

Leyland Leopard (1959-1982)

Leyland developed the Leopard single deck chassis in 1959 as a development to the Leyland Tiger Cub which had been launched six years earlier. Available as either a bus or coach chassis, the Leopard was initially powered by the O.600 engine, although later models had the more powerful 11.1-litre O.680 engine, both being underfloor and delivering a performance suitable for the newly opened motorway network. The early bus and coach chassis were both 30' long and 8' wide, although the 1961 amendment to Construction and Use Regulations introduced the 36' PSU3 version with pneumocyclic semi-automatic gearbox. A shorter version of the PSU3, the 31' PSU4, replaced the L1/L2 in 1965, and a second legislation change brought about the 39' long PSU5 in 1970.

The first L1 Leopards were delivered in August 1959 to Sheffield Joint Omnibus Committee, 1500-1505WJ being Weymann bodied coaches with 41 seats. Western SMT were not far behind in 1960, although their OCS725-744 had 30-seat Alexander coach bodies and were amongst the first vehicles ever fitted with a passenger toilet. The BET Group subsidiaries were the largest purchasers of the Leopards, using standard bodywork designs by Marshall, Weymann and Willowbrook for bus versions. Further north, the Scottish Bus Group acquired a significant bus fleet, mostly with Alexander Y-type bodywork, in Northern Ireland bodywork the Alexander X-type bodywork was preferred, and in the Irish Republic CIE chose to body many Leopards themselves. For Leopard coaches there was a wide variety of body choices available for selection by the operator, although Plaxton's Elite and Supreme or Duple's Dominant attracted the most orders.

Above: Maidstone & District moved away from AEC chassis in the mid-1960s, with their first order for Leyland Panthers soon followed by dual purpose Willowbrook bodied Leyland Leopard buses. 2801-2818 (OKO801G etc) were built in 1968 and entered service in 1969.

Typical of many Leopard operators was Southdown Motor Services, who were an early purchaser of the L2, ordering 30 distinctive Harrington Cavalier coaches with air suspension and just 28 seats in 1961, and a further 15 in 1962. Later deliveries were initially Plaxton Panorama I bodied for operating Southdown's long-distance stage services as well as private hire work (including the "21 Day Grand Tour of Britain") – many later succumbing to the anonymity of "National white". Southdown's first Leopard PSU3 buses for stage carriage work were delivered in 1967/8 and carried Willowbrook or Marshall 45-seat bodywork, with a batch of Northern Counties 49-seat dual-purpose buses following a year after. Barton Transport of Nottingham and Wallace Arnold Tours of Leeds were two of the largest Leopard coach operators, specifying both Duple and Plaxton coach bodies at various points during the Leopard's long production period. Towards the end of the Leopard's life, London Country purchased a pair in 1980 to evaluate against the Volvo B58: the Leopards "won" and resulted in 30 more being leased from Kirby Kingsforth in 1981 for a period of five years.

After a production run of in excess of 12,000 chassis, the Leopard was finally withdrawn from the Leyland range in 1982, with the final examples entering service in 1983. Amongst the last examples were East Lancs bodied buses for Rhymney Valley District Council, and numerically the last chassis was YEL98Y, an ECW bodied example delivered to Hants & Dorset in January 1983. Leyland had started to replace the Leopard with the new Leyland Tiger chassis in 1981.

Leyland Leopard Specifications

- L1 – 30' long bus (31' from 1961)
- L2 – 30' long coach (31' from 1961)
- PSU3 – 36' long, with pneumocyclic semi-automatic gearbox
- PSU4 – 36' long, with pneumocyclic semi-automatic gearbox
- PSU5 – 39' long

- add A B C D or E – to denote mechanical variations

- add /1R to /5R- to denote transmission and driving position

   - 1R – bus, synchromesh transmission, right-hand drive
   - 2R – bus, pneumocyclic transmission, right-hand drive
   - 3R – coach, synchromesh transmission, right-hand drive
   - 4R – coach, pneumocyclic transmission, right-hand drive
   - 5R – coach, ZF synchromesh, right-hand drive

Above: Safeguard of Guildford purchased this Leyland Leopard PSU3C/4R with Duple Dominant bus bodywork in 1976, one of eight similar buses bought between 1975 and 1983. It spent much of its life working Guildford local services "A" and "B" to the Park Barn Estate.

Daimler (Leyland) Fleetline (1960-1981)

Following close on the heels of Leyland who had introduced the Atlantean in 1958, Daimler's early Fleetlines were designed from the outset with a drop centre rear axle which allowed low height bodywork to be fitted. The prototype Fleetline (7000HP) was built with a Weymann body and a Daimler engine, but production models from 1962 were offered with a choice of Gardner's 6LW and 6LX units. Gardner's improved 6LXB engine was added in 1968, as was Leyland's O.680 in 1970. Originally offered as a standard 30' long chassis with an extended 33' option, a 36' version was added in 1968 based upon Daimler's existing Roadliner chassis, although with one exception this version was supplied solely for export customers. A single deck Fleetline chassis option was made available, but this found only a very limited number of customers in the UK.

Above: XDH516G is a 1969 Northern Counties bodied Daimler Fleetline, one of the final deliveries to Walsall Corporation before their merger into West Midlands PTE. Powered by a Gardner 6LX engine, it was the last former Walsall bus to be withdrawn by WMPTE.

From 1960, production of the Daimler Fleetline had been undertaken at Daimler's Coventry premises, although this relocated to Leyland's Farington factory in Lancashire in 1973, and the model was renamed the Leyland Fleetline in 1975. The last Daimler Fleetline was GWA836N, which joined South Yorkshire PTE in April 1975. Leyland developed a new version of the Fleetline, known as the B20, which received the more powerful Leyland O.690 engine and was modified to have improved engine compartment ventilation and lower noise levels. All B20 Fleetlines were provided to London Transport.

London Transport was by far the largest operator of the Fleetline, with 2,646 buses being used. Their decision to take the Fleetline over the Atlantean was made following comparison trials between 50 experimental Atlanteans ("XA") and 8 experimental Fleetlines ("XF") which commenced in 1965, with the Atlanteans carrying red for Central Area work (from Chalk Farm and Highgate garages), and the Fleetlines carrying green for Country Area work and being allocated to East Grinstead. The Fleetlines triumphed, and all 50 XF Atlanteans were sold in 1973 to China Motor Bus in Hong Kong. London Transport specified Park Royal or MCW bodywork for their Fleetlines, and the first (DMS1, EGP1J) entered service in January 1971. They were not popular with passengers, who noted that their frequent open platform Regents and Routemasters were being replaced with these larger buses on bus services which appeared to operate less frequently.

They were not popular at Aldenham Works either, when it was identified in 1976 that removing the Park Royal bodies for overhaul caused major distortion, and so the overhaul programme was cancelled and the expected lifespan of the Fleetlines dropped considerably. As the last of London's Fleetlines were arriving in 1978, so were the first of the new Leyland Titan (B15) "T" and MCW Metrobus "M" classes, and mass withdrawals of Fleetlines started in 1979.

Being based in Coventry, it is not surprising that Birmingham Transport ordered over 1,000 Daimler Fleetlines, bodied by MCCW or Park Royal, and also a small batch of single deck Fleetlines with 37-seat Marshall bodywork. The subsequent formation of West Midlands PTE in 1969 soon brought along over 1,000 more from other Midland fleets such as Coventry Corporation (East Lancs, Neepsend and Willowbrook bodied) and Walsall Corporation (Northern Counties bodied). Greater Manchester (SELNEC) PTE operated over 500 Fleetlines, including examples new to Manchester Corporation and others acquired with the fleets of Bury Corporation (Alexander and East Lancs bodied), Rochdale Corporation (MCW, Roe, Weymann and Willowbrook bodied), SHMD (Northern Counties bodied) to name but three. Other sizeable fleets were operated mostly by municipal fleets such as

Belfast, Cardiff, Derby and Sheffield, and THC/NBC operators including City of Oxford, Maidstone & District, Potteries, Southdown and Trent. Within Scotland, Alexander Fife, Alexander Midland, Central SMT, Scottish Omnibuses and Western SMT were notable Fleetline operators.

Hong Kong was a successful destination for the Fleetline, China Motor Bus receiving 336 and Kowloon Motor Bus 450 in the 1970s. As the London Fleetlines were withdrawn in the early 1980s, these provided a ready source of replacements for Hong Kong operators, and even once their days there were over, many were sold once again to operators in China. The complete list of international Fleetline orders was:

- CCFL of Lisbon (with CCFL bodies fitted), delivered 1967-72
- Johannesburg Transport (South African Bus Bodies), 1969-76
- Golden Arrow Bus Services (South African Bus Bodies), 1970-71
- Port Elizabeth Electric (with South African Bus Bodies), 1972
- CMB, Hong Kong (Alexander or Duple Metsec bodies), 1974-78
- KMB, Hong Kong (Baco or Duple Metsec bodies), 1974-78

The final year of Fleetline production was 1981, with Bournemouth Transport taking eight Alexander bodied examples and Southend Transport twelve built by Northern Counties. South Notts of Gotham received ECW bodied SCH116/7X, which were the very last Fleetlines. An unfortunate coincidence is that both the very first and very last Fleetlines were destroyed by fire: 7000HP in a bus depot fire in 1976, and then preserved SCH117X in a fire at the Nottingham Transport Heritage Centre in 2007.

- CRD6 – Coventry, Rear-engined, Daimler (six-cylinder engine)
- CRG6 – Coventry, Rear-engined, Gardner (six-cylinder engine)
- CRL6 – Coventry, Rear-engined, Leyland (six-cylinder engine)
- CRC6 – Coventry, Rear-engined, Cummins (six-cylinder engine)

BMMO CM6 & CM6T (1962-1966)

The Birmingham and Midland Motor Omnibus Company (BMMO) was a prominent bus operator in the Midlands Region between 1905 and 1981, although it was renamed to the Midland Red Omnibus Company in 1974. As well as being one of the largest bus companies in the UK, BMMO was also known for manufacturing its own range of buses and coaches between 1923 and 1969. Early models were known as "SOS" models, generally accepted as meaning "Superior Omnibus Specification", but from 1940 the vehicles were named "BMMO" and only made available for the company's own use.

Above: BMMO's CM6T - a fast "motorway express" coach of which 30 examples were produced 30 of the type. BHA656C entered service in 1965 on express services between Birmingham and Coventry and London and was fitted with coach bodywork for 44 passengers.

Aside from single and double deck buses, BMMO also constructed several coach models, and the CM6 and CM6T were designed as "Motorway Express" models for fast running to destinations such as London.

The M1 motorway opened in November 1959, and BMMO were quick to implement Motorway Express services between Birmingham, Coventry and London. Initially these workings were undertaken using re-engineered versions of the existing BMMO C5 coach, which had been fitted with 8-litre turbocharged engines, overdrive and on-board passenger toilet to become the CM5T class. The success of Motorway Express services soon required the development of a more powerful and larger coach, and prototype CM6T 5495HA (fleet number 5295) was constructed in September 1962, entering service from Digbeth Depot in Birmingham in March 1963. This 36' long coach was fitted with a huge BMMO 10.45-litre six-cylinder naturally aspirated engine connected to a five speed semi-automatic gearbox, easily capable of 80mph motorway performance (in the days before the motorways had a legal speed limit), and provided comfortable coach seating for 46 passengers. The CM6T was specified for high speed with servo assisted disc brakes fitted to both front and rear wheels.

Following minor modifications to 5295, BMMO commenced the production of a further 29 of the type for delivery in 1965 (5646 to 5663) and 1966 (5664 to 5674). 24 of them were to CM6T specification with 44 seats and an on-board passenger toilet, whilst five (5667 to 5671) were CM6 specification seating 46 without the on-board toilet. Painted in BMMO's coach livery of red bodywork and a black roof, the CM6Ts were mostly allocated to Birmingham or Coventry services to London, with the CM6s being allocated to the new M5 motorway services X43 and X44 from Birmingham to Worcester.

Being high mileage motorway coaches, it is not surprising that they were to be destined for a relatively short career, a fact not helped by unacceptable corrosion hastened by salt spread on motorways in icy conditions. Some were overhauled in the early 1970s to become CM6As (essentially a bus conversion involving removal of the toilet nd a repaint), but ultimately all the CM6 variants were withdrawn by 1974, with newer C14 class Leyland Leopards taking their place on motorway workings by then. Aside from the two survivors which remain today, actively preserved BHA656C (5656) and unrestored DHA962C (5662), all other CM6 variants were scrapped locally, all but one (5663) seeing no further passenger service.

Bristol RE (1962-1982)

The Bristol RE, meaning "Rear Engine" was a successful single deck bus and coach produced by Bristol Commercial Vehicles from 1962. The chassis was a basic design, fitted with leaf spring suspension, and with a rear, horizontally mounted engine at the rear. Unusually, the gearbox was located forward of the drop-centre rear axle to help weight distribution. A choice of six-cylinder engines were provided by Gardner (HLW/HLX/HLXB) or Leyland (O.680), and some models destined for New Zealand were fitted with the Leyland 510 engine, better known as the standard engine of the Leyland National. As with other Bristol chassis of the period, the majority were completed by Eastern Coach Works, and the bare chassis were driven from Bristol to Lowestoft for their bodied to be fitted.

A pair of prototypes were completed and released in 1962 (7431HN, B54F, supplied to United) and 1963 (521ABL, C47F, supplied to Thames Valley/South Midland). Initially the RE was only offered to subsidiary companies of the nationalised Transport Holding Company (THC), who also owned Bristol Commercial Vehicles. Following Leyland's acquisition of both Bristol and ECW in 1965, the RE was made available to British Electric Traction (BET) group companies, as well as municipal operators. This continued through 1970, with the formation of the National Bus Company, and only started to slow when the Leyland National was introduced in 1972. UK orders dried up by the end of 1975, and from 1976 until the end of production in 1982 the RE chassis was only built for Ulsterbus and Citybus in Northern Ireland, the largest operator of the type, and for the Christchurch Transport Board in New Zealand.

Above: Eastern Counties RLE684 is a 1970 Bristol RELL with an ECW bus body fitted with 50 dual-purpose coach seats. After 14 years' service it passed to Cambus in 1984 and had found its way to Citybus in Belfast as a driver training vehicle by 1988.

Whilst ECW was the most widely chosen bus bodywork option, an additional ten body manufacturers completed Bristol REs, including Alexander (Lincoln, North Western, Scottish Omnibuses, Western SMT), Alexander Belfast (Citybus, Ulsterbus), Duple (Hants & Dorset, Isle of Man, Southern Vectis), East Lancashire (Accrington, Halton, Widnes), Marshall (Aldershot & District, East Midland, North Western, South Wales, Western Welsh), MCW (Sunderland), Northern Counties (SHMD), Park Royal (Liverpool) and Plaxton (Greenslades, Lancashire United, Red & White, West Riding). The Christchurch examples were bodied locally in New Zealand by either Hawke or NZMB/Hess.

The final day of normal bus service for the Bristol RE was with Ulsterbus on 28th February 2004, although 33 of their Alexander (Belfast) bodied examples were acquired by Lough Swilly in Ireland for school duties until 2007. This era came to an end on 23rd June, when 84-DL-2348 (formerly Ulsterbus BXI2568) performed the last working. The Bristol RE enjoys popular support from bus preservationists, with many examples of both bus and coach variants being restored to their original operators' liveries.

Above: One of the last Bristol REs built was BXI2565, an Alexander (Belfast) bodied 52 seat RELL6G bus delivered to Citybus of Belfast in July 1983. It passed to Ulsterbus as fleet number 2565 in 2000 and was sold out of service in 2003. Whilst many similar Bristol REs were destroyed in the troubles in Northern Ireland, this is one of several examples that have been provided with a future in preservation.

Bristol RE Specifications

- RELL – low frame bus, long version (36')
- RELH – high frame coach/dual-purpose bus, long version (36')
- RESL – low frame bus, short version (33')
- RESH – high frame coach/dual-purpose bus, short version (33')
- REMH – high frame motorway coach, maximum length (36')

add engine type, e.g. RELL6L indicates a six-cylinder Leyland engine

AEC Renown (1962-1967)

AEC had been providing the Bridgemaster integral bus with Park Royal bodywork as competition to the Bristol Lodekka since 1957, and 179 models had been completed by the time production ended in 1962. Its replacement soon arrived in the shape of the AEC Renown, which although visually similar to the Bridgemaster was built as a separate chassis allowing operators a choice of conventional bodywork. It was powered by AEC's AV590 engine coupled to a monocontrol transmission, air suspension being provided to the Renown's rear axle, although having conventional leaf springs to the front. During its short production life, the Renown was bodied by East Lancs, MCW, Northern Counties, Park Royal, Roe and Willowbrook. Aside from Renowns delivered to Leicester City Transport and Leigh Corporation which had rear entrance bodywork, all others were fitted with front entrances. Two Renown demonstrators were produced in late 1962, with both 7552MX and 8071ML being bodied by Park Royal with a front entrance.

Significant users of the Renown included East Yorkshire Motor Services, Leicester City Transport, North Western Road Car, Nottingham City Transport, City of Oxford Motor Services, South Wales Transport and Western Welsh. London Transport borrowed 8071ML for trials in 1962/63, giving it fleet number "RX1". It was initially assessed at Aldenham before being moved to the Country Area's Northfleet garage for four months of running on the busy 480 service. It was returned to AEC in August 1963 and was sold on to Osbournes of Tollesbury by the end of the year. London Transport did not place any orders for the Renown. Of eight AEC Renowns purchased new by independent operators, King Alfred Motor Services of Winchester received 595/596LCG in 1964, which subsequently passed to Hants & Dorset as 2211/2 in April 1973.

Above: King Alfred Motor Services of Winchester received two AEC Renowns with Park Royal 75 seat bodywork in June 1964, 595LCG and 596LCG. These later passed to Hants & Dorset Motor Services (as 2211 and 2212) in April 1973 when the business was purchased. Both are now preserved by the "Friends of King Alfred Buses".

The last of 252 Renowns was produced in 1967, with many of its previous customers now choosing the new rear-engined Leyland Atlantean and Daimler Fleetline chassis for their future needs. The last completed example was FWL371E, one of four Northern Counties bodied Renowns delivered to City of Oxford Motor Services in August 1967.

Bedford VAL (1963-1973)

Bedford Vehicles launched their VAL coach at the 1962 Commercial Motor Show, and production ran for ten years between 1963 and 1973. It was a visually distinctive vehicle, having two front steering axles in a "Chinese Six" format, allowing Bedford to develop a chassis to the new maximum length of 36' whilst continuing to use existing stock axles. The original Bedford VALs were fitted with a six-cylinder 6.2-litre Leyland diesel engine and designated "VAL14", although a new version in 1967 introduced the larger (and somewhat quieter) 7.6-litre Bedford engine, with the new model designated "VAL70". The VAL chassis found a ready following, at just £1,775 for the VAL14 and £1,910 for the VAL70 it was a significantly cheaper option than the equivalent length AEC or Leyland chassis available at the time and being over a tonne lighter returned a very respectable 15 mpg. The dual front wheels were marketed as a safety feature for the faster speeds of the UK's developing motorway network: if one front tyre failed then the vehicle could still be brought to a safe stop.

The majority of Bedford VAL14s were built as coaches with 52-seat bodywork supplied by Duple or Plaxton. The 1967 launch of the VAL70 aligned with both Plaxton's launch of their new Elite bodywork range and Duple's unveiling of their new Viceroy body, both seating 53 passengers. A smaller number of coach-bodied examples were also supplied by Harrington, Willowbrook and Yeates. The Portuguese manufacturer Caetano offered its Estoril body, although very few were actually ordered, and VanHool bodied just one with its unusual Vistadome bodywork in 1971 (RAR690J). Both versions of the VAL chassis were a popular choice for UK coach operators, from the large national fleets such as Wallace Arnold of Leeds, who operated a mixture of Duple and Plaxton bodied coaches, to the smaller independents such as King Alfred Buses of Winchester who ordered just two Plaxton examples (CCG704C and EOU703D). The VAL was, however, an unusual choice for bus work, with 20 being bodied by Marshall with centre and rear doors for airside work with British European Airways (BEA) at Heathrow Airport (LMG155-164C, OGO 337/340E and OYF262-269F). A further 10 were bodied by Strachans of Hamble for North Western Road Car's Altrincham depot, these lowbridge single deck buses with their rounded roof profiles were ideal for passing under the Dunham Massey canal bridge.

Above: Abbey Coachways of Selby operated a fleet of 13 Bedford VALs, three new VAL14s, three new VAL70s and seven acquired second hand from other operators. WWY115G is a VAL70 with Plaxton Panorama Elite bodywork, delivered new in January 1969. It was retired in 1976 becoming staff transport for a paint company.

The Bedford VAL is perhaps best remembered for its high-profile movie career. 1964 Harrington Legionnaire bodied ALR453B was the star of "The Italian Job", which saw it transporting the famous trio of Mini Coppers before ultimately coming to rest hanging precariously over the edge of an Italian mountain pass, being pulled over by the weight of stolen gold. It was new to Battens Coaches of East Ham in 1964, purchased and modified by Paramount Films in 1968, and as well as being an important part of the film plot was also used to transfer the crew and their equipment to and from the various Italian filming locations. After filming had been completed, it was converted back to a standard coach and saw further PSV usage with Wendy's Coaches of Liverpool. Another film star was 1967 Plaxton Panorama I bodied VAL14 URO913E, new to Fox Coaches of Hayes in Middlesex, which was chartered by EMI to be featured in the Beatles film the "Magical Mystery Tour". It is still thought to exist in storage with the Hard Rock Café in the USA, having been noted at their restaurants in both Florida and California. Until recently three other Plaxton bodied Bedford VAL70s operated the Beatles Magical Mystery Tour in Liverpool, including DJH731F, OOR320G and SCK56K.

Over 2,000 Bedford VALs were produced in a little over 10 years. Whilst many enjoyed second lives as car transporters or caravans, an increasing number have been rescued and undergone extensive restoration by preservationists. It is worth noting that some Bedford VALs were built new as non-PSV vehicles and some still survive, including examples of caravans, horseboxes and even an outside broadcast unit (OOW999G), which was new to Southern Television in Southampton in 1968.

AEC Swift / Merlin (1964-1974)

Following the popular trend of placing bus engines at the rear of the chassis, AEC of Southall introduced the single deck Swift chassis in 1964, which was available in two lengths (33' or 36') and with either the AEC AH505 or AH691 engines. Being available for bodying by a variety of companies, both in single and dual door formats, it gained popularity with operators as it was suitable for one-man operation at a time when labour shortages impacted upon the provision of some bus services. The first demonstrator was completed in November 1965, being a 53-seat Willowbrook bodied Swift (FGW498C).

London Transport (and London Country Bus Services) were the biggest single purchasers of the AEC Swift, taking more than 1,500 between 1966 and 1972 with similar dual door bodywork manufactured by Marshall of Cambridge, MCW, Park Royal and Strachans. Early deliveries were of 36' Swifts, which were uniquely known as "Merlins" in London, although their length introduced operational issues in London's busy traffic, so the shorter 33' Swift was selected instead between 1970 and 1972. They were designed to be "crowd shifters" with more standee capacity than passenger seating, and many were initially fitted with automatic fare collection equipment and turnstiles in an attempt to shorten loading times. Six Merlins were the first buses to be used on the new high frequency limited stop Red Arrow services introduced in April 1966 between Victoria and Marble Arch, and more of the type were allocated to this role as additional Red Arrow services were introduced over the next few years.

Above: AML11H was a 1970 AEC Swift with Marshall bodywork delivered to London Transport as fleet number SM11. After ten years' service it was withdrawn and sold to Malta in 1981, where it continued to be used on a daily basis for the next 30 years, only being withdrawn when Arriva Malta took over the island's bus services in July 2011.

The Merlins and Swifts were not assessed as being overly successful buses for London and combined with new legislation which permitted one-man operation of double deck buses lead to a formal disposal programme commencing in 1972. The Red Arrow Merlins survived until 1981, when they were replaced by new Leyland National 2s.

The AEC Swift was selected by several municipal fleets, including Belfast Corporation, Birmingham City Transport, Blackpool Transport, Cardiff Transport, Great Yarmouth Transport, Leeds City Transport, Merseyside PTE, SELNEC, Sunderland Corporation and Tyne & Wear PTE, although the majority of these had been withdrawn and sold by the early 1980s. Hundreds of AEC Swifts were purchased by two Australian operators, the Department of the Interior and MTT Adelaide and were joined by even more former London examples as these became available for purchase. AEC Swift production ended in December 1975, with the final three UK models being Marshall bodied JEE48-50P for the Grimsby Cleethorpes fleet. 177 second-hand Swifts and Merlins were purchased by Citybus in Northern Ireland between 1977 and 1980 to replace the many buses that had been destroyed in the riots.

At least 60 AEC Swifts were exported to Malta from London Transport, most being used as route buses, but a small number were allocated to the Education Department for school transport. Whilst some remained easily recognisable as former London Transport buses, others were subject to extensive modifications which made their identification somewhat more difficult. Many continued to operate on the Mediterranean island on a daily basis until Arriva Malta took over the public transport system in July 2011.

Bristol VR (1966-1981)

Bristol designed the VR chassis in 1966 as a planned replacement for its Lodekka range, the "VR" indicating that the engine was vertically mounted at the offside rear of the vehicle. A choice of three engines were made available, Gardner's 6LW and 6LX and Leyland's O.600, all driven by a semi-automatic gearbox, and the chassis was available in 10 and 11 metre lengths. Bristol completed two prototypes in time for the 1966 Earls Court Motor Show: GGM431D and HHW933D both had 80-seat bodywork built by Eastern Coach Works of Lowestoft. GGM431D subsequently joined the Central SMT fleet, and HHW933D joined Bristol Omnibus, but spent much of its early years further north with Mansfield District Traction Company. In 1967, Bristol modified the design by changing to a transverse engine (VRT), which made it eligible to receive the 1968 Government Bus Grant brought in to encourage fleet modernisation and one-man operation.

Above: CPU979G, fleet number 3000 in the Eastern National fleet, was the first Bristol VRT/SL, and its chassis had been displayed at the 1968 Earls Court Show. It was in passenger service between 1969 and 1986, when it became a driver trainer until withdrawal in 1991.

Production commenced in 1968, with the first VRT/LL (Long, Lowbridge) examples joining the Scottish Omnibuses fleet in November and the first VRT/SL (Short, Lowbridge) models joining Eastern

National (CPU979G) and Brighton, Hove & District (OCD763G) in March 1969. The Scottish VRs soon became troubled by engine overheating and gearbox issues, which ultimately resulted in Central SMT, Eastern Scottish and Western SMT "swapping" 91 VRTs for late 1967/68 Bristol Lodekkas from various National Bus Company subsidiaries. Bristol worked to improve these matters, and introduced the Series 2 VR in 1970, externally visible by its three-piece engine cover instead of the one piece fitted to the Series 1, and the first Series 2 (OFC901-3H) were delivered to City of Oxford in April 1970. The Series 3 VR followed in 1974 (prototype only, normal production commenced in August 1975), and was designed to ensure compliance with updated engine noise regulation. The short low-height version of this VR ("VRT/SL3") soon became the standard double deck vehicle for many National Bus Company subsidiaries.

The majority of Bristol VRs were completed by ECW, which provided a variety of height options between 4.09m and 4.42m, including a convertible open-top version at 4.22m. The standard bus configuration was for 74 seated passengers with a single front door, being reduced to 70 for the dual-door configuration. Other companies building on the Bristol VR chassis included Alexander (Cardiff, Northampton), East Lancashire (Burnley & Pendle, Lincoln, Merseyside PTE, Northampton, Sheffield), Northern Counties (Cleveland Transit, Gelligaer, Reading), MCW (West Midlands PTE) and Willowbrook (Cardiff, East Kent, Northern General). Two batches were exported, both being bodied by Bus Bodies (South Africa) Ltd, for Johannesburg in1969 and Pretoria in 1971.

The Ribble subsidiary fleet of Standerwick required a new double deck coach for motorway journeys at the end of the 1960s, to replace its older "Gay Hostess" Leyland Atlanteans on services between the North West, the Midlands and London. A Bristol VR coach was developed based upon the VRL/LH (Long, Highbridge) chassis, powered by a Leyland 680 "Power Plus" engine connected to a five speed semi-automatic gearbox and fitted with power steering. A prototype with fleet number 50 (FCK450G) was delivered in 1969, and despite its low clearance being reported as having caused damage to the road surface at several bus stations in the North West, another 29 followed over the next three years. The 60-seat ECW bodywork was designed with a centre entrance door, a staircase towards the rear, a large luggage boot, a passenger toilet and three skylights in the roof of the upper deck.

Above: Ribble subsidiary Standerwick ordered 30 Bristol VRL/LH double deck coaches at the end of the 1960s to replace older Leyland Atlanteans on "Gay Hostess" motorway express services. Bodied by ECW with 60 coach seats, LRN60J was delivered to Standerwick in 1971, passed to National Travel North West in 1974 and was withdrawn and sold in 1976 after only five years in service.

The last of 4,531 Bristol VRs was completed in 1981, and the final models delivered to Bristol Omnibus and Southdown were different in being fitted with Leyland's 680 engine. The Bristol VR was retired and replaced by the Leyland Olympian, which was also the planned successor to the similarly retired Leyland Atlantean chassis. Withdrawn VRs were a popular choice for other operators, many continuing in daily service as sightseeing buses or school transport well into the twenty-first century. A number went abroad, their lowbridge bodies suiting the lower bridge clearances to be found in mainland Europe. A small number achieved media fame: one was used in the Spice World film in 1998, and three were used by Dr Who in an episode in 2009. The Bristol VR is an increasingly common sight at bus rallies and events, with privately preserved vehicles showing the many different vehicle configurations and liveries carried by its many operators.

<u>Bristol VR Specifications</u>

- VRX – VR prototypes (GGM431D and HHW933D)
- VRL/LH – Series 1, long, highbridge
- VRT/LH – Series 1, long, highbridge
- VRT/LL – Series 1, long, lowbridge (for Scottish Omnibus)
- VRT/SL – Series 1, short, lowbridge
- VRT/LH2 – Series 2, long, highbridge
- VRT/LL2 – Series 2, long, lowbridge (for Reading Transport)
- VRT/SL2 – Series 2, short, lowbridge
- VRT/LL3 – Series 3, long, lowbridge
- VRT/SL3 – Series 3, short, lowbridge

add engine type, e.g. VRT/SL3/6G is a six-cylinder Gardner engine

Bristol LH (1967-1981)

Bristol Commercial Vehicles designed and produced the Bristol LH chassis for bus operators who needed a smaller, lightweight vehicle, often for the operation of rural services, the "LH" indicating a <u>L</u>ightweight chassis with a <u>H</u>orizontal engine. The LH was available in three lengths: the standard LH was 9.1 metres long, the shorter LH<u>S</u> version only 7.9 metres long, and the longer LH<u>L</u> option was 11 metres in length. The LH was initially constructed to be 2.3m wide, with a wider 2.5m version being added later. Operators chose between two six-cylinder diesel engine options: Leyland's 6.5-litre O.400 (and later O.401) producing 138 bhp or the smaller 5.8-litre Perkins H6.354 which produced 101 bhp.

Available to be bodied as either buses or coaches, the vast majority of buses were completed by Eastern Coach Works of Lowestoft whilst coaches were most commonly bodied by Plaxton of Scarborough. Other LHs were bodied by Alexander, Duple, East Lancashire, Marshall, Northern Counties, Weymann and Willowbrook. Seating format varied according to the length of the LH and whether it was completed as a bus or coach: the standard LH normally accommodated 31 coach seats or 41 bus seats, the shorter LHS offered either 26 or 35 seats, and the longer LHL 53 or 55 seats.

Above: The LH6L chassis was suitable for narrow coaches, and this 1975 Greenslades example has a Plaxton Panorama Elite III body for 45 passengers. Being 7'6" wide, it was ideal for Dartmoor Tours in Devon. JFJ506N also worked for a while on Guernsey.

Three LH prototypes were built in 1968 and 1969, being NHU100F (an ECW bodied 45-seat bus with a Perkins engine), CJJ44H (a Plaxton bodied 49-seat bus with a Leyland engine) and MBO1F (a short LHS model with 30-seat Weymann dual-purpose body which had originally fitted to Albion Nimbus WKG27). The standard length bus-specification LH became popular with a number of National Bus Company subsidiaries and some municipal fleets, with the largest orders being received from Bristol Omnibus (116 LH), Crosville (56 LH), Eastern Counties (65 LH, 5 LHS), Hants & Dorset (120+ LH), Lincolnshire Road Car (96 LH, 10 LHS), London Country (67 LHS), London Transport (95 LH, 17 LHS), United Automobile Services (218 LH and 5 coaches) and Western National (120 LH, 71 coaches and 18 LHS). Within Scotland, Alexander Y-type bodywork was preferred, with Alexander Midland ordering 3 buses and 38 coaches and Eastern Scottish a further 34 coaches, and Alexander (Belfast) bodywork was carried by 100 Bristol LH buses delivered to Ulsterbus in Northern Ireland in 1973. Of a total of 1,987 Bristol LHs produced, well over half were used by the fleets noted above.

The Bristol LH was succeeded by the lighter Leyland National "B" which was introduced in 1978, and the final Bristol LH was CLJ413Y, a 35-seat Plaxton bodied bus delivered to Bere Regis & District in March 1983. The LH enjoyed a varied second life with independent operators, and a large batch was exported by Ensign to become route buses on the Mediterranean islands of Malta and Gozo. LHs from Crosville, Hants & Dorset, Southern National and Western National remained in daily use on the island until their withdrawal following the July 2011 takeover by Arriva Malta, although at least one (GLJ458N) has passed to Heritage Malta for preservation. Former London Transport OJD91R was unusually converted to open top format in 2010 to provide Jersey island tours from St Helier. An increasing number or Bristol LH buses and coaches are being privately preserved, including prototype MBO1F.

Bristol LH Specifications

- LHS – Bristol LH Short (26')
- LH – Bristol LH standard length (30')
- LHL – Bristol LH Long (36')

Bedford Y-Series (1970-1986)

Bedford's Y-Series of coach chassis was launched in September 1970 and continued to be offered until the end of production at their Dunstable factory in 1987. Each different model was given a three-letter code: the initial "Y" denoting the mid-engined underfloor chassis, the second letter indicating the engine type, and the final letter the vehicle's length (e.g. Q = 10m, T = 11m, V = 12m). The first Y-Series model was the YRQ, powered by Bedford's 466 cubic inch engine connected to a Turner gearbox, although by 1974 this has been replaced with the Eaton gearbox, shared with the longer 12 metre YRT chassis. Bedford's upgraded 500 cubic inch engine became available in 1975, with the YRQ being updated to become the YLQ, and the YRT becoming the YMT. A turbocharged Bedford "Blue Series" engine was added to the range in 1980 as the YNT, although the existing YLQ and YMT chassis were offered with a non-turbocharged version of the same engine.

Mostly favoured by independent coach operators, the Y-Series were commonly bodied by Duple or Plaxton although many were bodied by Willowbrook as service buses. Some examples found their way into larger fleets: South Wales Transport ordered 18 Duple Dominant bodied YMQ buses in 1981 (FCY280-297W). Boro'line Maidstone standardised on various Bedford Y-Series chassis for their bus fleet following their decision to discontinue double deck bus operations in 1977: a snapshot of their 1980 fleet illustrates the use of seven different chassis and bodywork combinations in just this one municipal fleet:

- Bedford YLQ, YMT, YRQ and YRT with Duple Dominant bus body
- Bedford YMT with Duple Dominant II Express coach body
- Bedford YRT, with Duple Dominant Express coach body
- Bedford YRT, with Willowbrook 001 bus body

1981 saw the introduction of the short YMQS chassis, powered by Bedford's 500-series 8.2-litre engine connected to an Allison automatic transmission. Most were bodied as service buses using the comparatively rare Lex Maxetta 33 or 37 seat-body, although Wadham Stringer also produced bus versions and Duple's Dominant was the most common coach version. United Counties ordered three Lex examples (WNH50-52W) for a network of midibus services in Leighton Buzzard, South Wales Transport introduced five Lex buses (LCY298-302X) to replace its short AEC Regent Vs in 1982 and Boro'line Maidstone took just one (CKN332Y) from dealer stock. Eastern National operated ten with Wadham Stringer bodies and dual-purpose seating for 33 passengers for rural services (TJN973-982W). All the South Wales examples soon passed to Alder Valley, who purchased two more (B918/919NPC), and many of these ultimately found their way into the rural fleet of Tillingbourne of Cranleigh.

Above: New in April 1976 to Florence Motors of Morecambe as SBV 284P, this Bedford YRQ carried Plaxton Supreme III bodywork for 45 passengers. By 1979 it had passed to Tours on the Isle of Man, becoming A111MAN, and was later was sold again to become a Route Bus on Malta, originally being registered Y-0475.

The last example to be completed was Bedford YNT E66JJU in May 1988, which was delivered to Grant & McAllin Coaches of Sheffield. Many surplus Bedford Y-Series were purchased by operators on the Mediterranean islands of Malta and Gozo, where they continued to provide reliable transport as route buses until their withdrawal in July 2011.

Leyland National (1972-1985)

The National Bus Company, formed in January 1969, worked with British Leyland to produce the Leyland National single deck bus to replace all the rear-engined single deck Leyland vehicles being produced at the time. Constructed within a specially built facility in the Cumbrian town of Workington, the Leyland National was a rear-engined integral vehicle of modular construction, allowing for maintenance and repairs by the operator themselves. It was powered by an 8.3-litre turbocharged Leyland 510 engine. The Leyland National was designed primarily for the National Bus Company, and from 1972 its subsidiary companies started to operate hundreds of examples mainly configured as single-door buses, although some operators such as London Country specified coach seated examples for long-distance services. The National was offered in plain white or National Bus Company green or red. It soon became a popular vehicle choice away from the National Bus Company, with London Transport and several municipal fleets building their own sizeable fleets. London Transport eventually ordered over 500 Leyland Nationals (both Mark 1 and Mark 2), and as a result London's shade or red was also added as a standard Leyland National colour.

A "Suburban Express" demonstrator (RRM148M) was unveiled in 1973, which was essentially a high floor National with coach seating on one level, instead of having the high step towards the rear of conventional Nationals. It toured the country as a demonstrator for a few years before passing to West Midlands Police in 1977, and later Suffolk County Council in 1984 for school transport, before being privately preserved. Another 1973 demonstrator was the Leyland National "Executive Commuter" (UTJ595M), fitted with just twenty aircraft style seats, colour televisions, tables and a bar. Some seats were designated as "super seats" equipped as mini offices with a range of audio-visual functions and full access to the secretarial service located at the front of the bus, which included the option of making calls by radio telephone. Not surprisingly, it did not attract any orders.

Above: Leyland produced a single "Suburban Express" version of the Leyland National in 1973, which was built as a high floor bus at one level throughout. It was initially displayed at Earls Court in 1973, and then became a demonstrator to fleets throughout the UK for several years. It then spent time with West Midlands Police and then Suffolk County Council, before being acquired for preservation.

A "Suburban Express" demonstrator (RRM148M) was unveiled in 1973, which was essentially a high floor National with coach seating on one level, instead of having the high step towards the rear of conventional Nationals. It toured the country as a demonstrator for a few years before passing to West Midlands Police in 1977, and later Suffolk County

Council in 1984 for school transport, before being privately preserved. Another 1973 demonstrator was the Leyland National "Executive Commuter" (UTJ595M), fitted with just twenty aircraft style seats, colour televisions, tables and a bar. Some seats were designated as "super seats" equipped as mini offices with a range of audio-visual functions and full access to the secretarial service located at the front of the bus, which included the option of making calls by radio telephone. Not surprisingly, it did not attract any orders.

Above: KPA369P is a 1975 Mark 1 Leyland National delivered new to Alder Valley (218). After withdrawal it passed to British Bus Sales of Cobham and was sold for preservation in June 2003, having been assessed as being too good to scrap. 218 changed hands again in October 2010, joining the fleet of Mortons Travel of Tadley.

The standard Leyland National was offered in two standard lengths of 10.3 and 11.3 metres, with single or dual door options. Buses built between 1972 and 1978 featured the distinctive roof mounted heating pod which produced warm air at roof level: in 1978 a "B" version was introduced that did away with the roof pod and introduced under-seat heating reducing both cost and weight. In 1979, the Leyland National 2 was introduced, with a revised front end incorporating the radiator and a wider choice of engines at the rear. The Leyland 510 was replaced by the 11.1-litre 0680 and then the L11 engine, and by 1982 was being offered with the popular Gardner 6HLXB or 6HLXCT units. Aside from these changes, the Leyland National 2 was visibly very similar to the National 1.

From 1979, the National 2 was used as the basis for the Danish-built Leyland-DAB articulated bus, which was powered by Leyland's O.690 engine and a ZF transmission. These sold well within Denmark, but only received limited interest from UK fleets. South Yorkshire PTE ordered two batches, the initial five having Leyland National 60-seat bodywork and were used as part of a government authorised experiment into future travel needs, whilst the second batch of 13 looked visibly different as they carried DAB bodywork. British Airways at London's Heathrow Airport purchased seven in 1981, these were bodied by Roe for airside duties, with five sets of doors (three nearside and two offside), and a capacity of 45 seated passengers and 100 standing.

The Leyland National 2 continued to sell well until National production ended in 1985, when it was replaced by the new Leyland Lynx. Following trials with a National 2 loaned from Ribble, London Transport became the largest operator of the type when an order for 69 National 2s (GUW438-506W) arriving in 1981 to replace the elderly AEC Merlins that were still operating their high frequency Red Arrow services. These were dual door versions with just 24 seats and a standing capacity for 46 passengers, performing well enough that 41 were considered suitable for "Greenway" conversion between 1992 and 1994 (see below).

Leyland Nationals were subject to modifications and rebuilds in their later lives, the most common being the replacement of the original Leyland engine with newer lower emission DAF or Volvo units. East Lancashire Coachbuilders went further in developing the "Greenway" mid-life rebuild programme for the many Nationals still in use by the

London & Country fleet. Donor Nationals were stripped back to their chassis and body frame and received a new Gardner engine at London & Country's Reigate garage, then were transported to East Lancs at Blackburn for re-panelling, new glazing and replacement front and rear end panels. The first completed Greenway, JCK852W, entered service with London & Country in 1991, and over 175 were subsequently converted. The majority were produced for the London & Country, Kentish Bus, London General (Red Arrow) and Crosville Wales fleets, although the Crosville examples were acquired by Luton & District (The Shires) within two years.

Above: new to London Country Bus Services, NPD142L worked until withdrawal in 1983. It later ended up with Midland Fox in 1989, being selected to be part of the Greenway conversion programme in 1994.

Having produced well over 7,000 Leyland Nationals, preserved examples are becoming an increasingly common sight at bus rallies and events around the country.

Diversion 1: in an effort to demonstrate the versatility of the Leyland National, the "Leyland Lifeliner Casualty Unit" was created and displayed at the 1974 Commercial Motor Show. Registered LSN3N, it was a large combined ambulance and disaster response vehicle, with publicity claiming that it would help stabilise injured persons at the scene before transferring them to hospital in conventional ambulances. It was fitted with a patient processing area, a rear patient stretcher lift, data processing capabilities, CCTV, floodlights, internal and external environmental monitoring and an 8-ton winch on the front bumper. No orders ever came, and LSN3N was converted back to a standard Leyland National bus in 1975, becoming GOL438N in the Midland Red fleet.

Diversion 2: to complete the story of the Leyland National, mention must be made of the two classes of "Railbus" train built for British Rail in the 1980s which used Leyland National bus components. The "Pacer" series was developed jointly by British Rail's Research Division and Leyland Motors in the mid-1980s, and the two-car unit comprised National bus bodies (with modified front and rear panels) mounted on a rail freight vehicle frame and powered by a Leyland engine. British Leyland later constructed the British Rail Class 155 "Super Sprinter" at its Workington factory in 1987/88, again using many Leyland National components.

Leyland National Specifications

- First set of characters is the bus length – e.g. 103 = 10.3 metres
- Second set of characters is the engine – e.g. 51 = Leyland 510
- Single letter: A = first series, B = Series B, H = Leyland National 2
- Doors and driving position – e.g. 2R = two doors, right-hand drive

(e.g.10351B/2R is a Series B, Leyland 510-engined, 10.3-metre-long, dual-door and right-hand drive bus)

Ford Transit (c.1972-c.1988)

Ford introduced their Transit range of vans, minibuses and pickup trucks in October 1965. London Transport trialed Dormobile bodied Ford Transit demonstrator JUD103L in 1972 in areas of London which were not served by existing routes, perhaps due to narrow roads or other access issues. 20 Strachans bodied Transits (FS1-20, MLK701-720L) with 16 seats entered service later in 1972 on routes B1, C11, P4 and W9, and after several years all four routes were assessed as having been successful enough for them to be upgraded with newer Bristol LH buses. Early Transit buses had a bodywork style which was visually similar to the shape used for mobile shops in the early 1970s, which gained them the nickname "bread vans". 1974 saw the introduction of London Dial-a-Ride services with the H2 between Golders Green Station and Hampstead Garden Suburb and the PB1 from Potters Bar to Rushfield, and an additional Ford Transit was acquired in 1975 (FS21, GHM721N) with five more in 1979 (FS22-26, CYT22-26V). Three final Carlyle bodied Transits were added to the H2 Dial-a-Ride in 1989 (FS27-29, C502/3/1HOE), these having a larger capacity for 20 seated passengers.

Whilst most Transit activity was taking place in the Capital, there were isolated pockets of experimentation taking place elsewhere in the UK. London Country trialed a Dial-a-Ride service in Harlow in 1974 using five Dormobile bodied 16-seat Transits (FT1-5, VPF121/2M, XPE123-5N), and the success of this route also resulted in the allocation of newly delivered Bristol LH buses in 1977. In the early 1980s, several operators realised that they could operate small minibuses such as the Ford Transit more cheaply than full sized service buses, access housing estates and other areas which were unsuitable for their larger vehicles, and recruitment challenges would be addressed as minibus drivers needed less stringent driving license classes. The new Manager of Devon General, Harry Blundred, arranged to evaluate a Ford Transit minibus in late 1983, and had decided by February 1984 to introduce a high frequency minibus network in Exeter using 16-seat Carlyle bodied Ford Transits. Within two years Devon General had more than 200 of the type in service, and Ford Transit operations had been extended to other Devon towns such as Torbay ("Bayline") and Teignmouth ("Teignbus").

Above: Ford Transit C760FFJ was delivered to Devon General in 1986, one of over 100 similar Carlyle- bodied minibuses to enter service that year. 760 was allocated to the Exeter network, and is now preserved in a livery similar to its original scheme.

Encouraged by the successful results being obtained in Devon, other operators started to experiment with minibus services, and in 1985-86 the Ford Transit was added to many major fleets, including Bristol, Cambus, Eastern Counties, Eastern National, East Yorkshire, Hampshire Bus, South Midland, Southern Vectis, Thames Valley & Aldershot, United Counties and West Midlands PTE. The majority were bodied by Carlyle in Birmingham, and amongst the last examples produced by them in late 1986 were Transits for Alder Valley North, Alder Valley South and Shamrock & Rambler. Mellor of Rochdale were also a popular choice to produce minibus bodies for Transits, particularly for Thames Transit, and these continued to be produced until the late 1980s.

The deregulation of the bus industry in 1986 also presented the opportunity for minibuses such as the Ford Transit to be used cost effectively in competitive markets. Harry Blundred launched Thames Transit in March 1987 in Oxfordshire, using buses from Devon General (including Mellor bodied Transits) to compete with City of Oxford Motor Services by providing high frequency services. Oxford responded by introducing their own Ford Transits, branded as "City Nippers", and one example (C724JJO) has been preserved in this livery within the nearby Oxford Bus Museum.

With the value of small bus operation having been proven, operators started to identify better alternatives to replace the Ford Transit, with Freight Rover Sherpas, Ivecos and Mercedes all gaining a hold in the UK minibus market as a result.

Seddon Pennine 7 (1974-1982)

In 1970, Leyland announced that it was to end production of the manual gearbox Leyland Leopard, which concerned the Scottish Bus Group which was not keen on purchasing vehicles with semi-automatic gearboxes. Instead, they contacted Seddon Atkinson, who were at the time working with Ailsa on a new double deck chassis, and after swift development a prototype single deck chassis named the "Seddon Pennine 7" was produced in October 1973.

The Pennine 7 chassis was available in two lengths: 11m (5.6m wheelbase) and 12m (6m wheelbase) and was powered by a horizontally mounted Gardner 6HLXB engine. Whilst the prototype had a semi-automatic gearbox, production Seddon Pennine 7s were supplied with a four-speed ZF manual gearbox with synchromesh, although different final ratio configurations provided a top speed of 56mph for service buses and 76mph for coaches.

Above: ESC847S is a 1978 Seddon Pennine 7 with Alexander Y-Type bus bodywork, fitted with 53 dual- purpose seats. It was new to Eastern Scottish in Edinburgh as S847A, and later passed to Lowland Scottish in 1986 for a further two years' service.

Whist the chassis was effectively produced for the Scottish Bus Group, only two of its subsidiary companies purchased the Seddon Pennine 7, Eastern Scottish and Western SMT. Alexander provided 53-seat bus bodywork or 45-49 seat coach bodywork for the majority of the chassis, with the exception of 64 Plaxton Supreme coach bodies ordered by Eastern Scottish. In 1974, Eastern Scottish progressed to order the longer 12-metre Pennine 7 chassis to carry the luxurious Alexander M-type coach bodywork which was favoured for long distance coach services from different Scottish cities to London. Whilst other Scottish operators chose to place this style of coach bodywork on the Volvo B58-61 or Leyland Leopard chassis, Eastern Scottish chose the new Seddon Pennine 7, which was supplied with a six-speed ZF manual gearbox with synchromesh, and a top speed of over 85mph. When these finally arrived in 1975, they carried a new blue and white Scottish Bus Group livery which would become the standard colour scheme for the next ten years.

Both the 11 and 12-metre versions were soon attracting orders from smaller, independent coach operators, which were bodied almost exclusively with Plaxton Supreme bodywork. Bowing to pressure, Leyland started to offer manual gearbox options on its Leopard chassis to Scottish Bus Group companies, the ready take up of which resulted in a reduced demand for the Pennine 7: by the time the new Leyland Tiger had been launched production was all but complete. The Seddon Pennine 7 was retired in 1982, after a total of 527 chassis had been completed. The very last example was Willowbrook 003 bodied BNC334Y which joined the fleet of Johnson of Oldham, and which coincidentally was also the last 003 body to be completed before Willowbrook closed in 1983 following legal action taken by the National Bus Company. Second hand Pennine 7s were not very popular, although some examples did find a second hold with fleets such as Stevensons of Uttoxeter and North Eastern Bus Service of Newcastle.

Fortunately, several Seddon Pennine 7s remain actively preserved, including several bus examples. Perhaps unique in preservation is MSF750P, one of the Alexander M-type 42-seat motorway coaches delivered to Eastern Scottish in 1975.

Dennis Dominator (1977-1996)

Dennis produced their first rear-engined double deck bus chassis in 1977, ending a ten-year absence since production of the Dennis Loline completed in 1966. Built as a 9.5 or 10.3-metre-long chassis, the Dominator was available with many different options which provided great variety over its 20-year production run. The default engine was the Gardner 6LXB connected to a Voith DIWA gearbox, although Cummins L10, DAF, Rolls Royce Eagle and other Gardner engines were also offered. Bodywork was mostly by Alexander, East Lancashire, Marshall, Northern Counties or Willowbrook.

The Dominator was also produced as a single deck chassis at the request of Darlington Corporation, and Marshall dual door bodied YVN73T was demonstrated at the 1978 NEC Motor Show in Birmingham. Whilst Darlington was the largest purchaser of the single deck version, Merthyr Tydfil ordered six in 1979 and Thamesdown Transport four short wheelbase buses in 1980. Barrow and Hartlepool both bought a small number of single deck Dominators, bur theirs were East Lancashire bodied examples. The most unusual single deck Dennis Dominator was EBB846W, which was supplied new to fire engine builder Angloco in 1980 to become a mobile control unit for the Tyne & Wear Metropolitan Fire Brigade.

South Yorkshire PTE operated one of the first production Dennis Dominators, SHE722S joining their fleet in November 1977, and grew to become the biggest user of the Dominator with 323 examples working until final withdrawal in the summer of 2006. Leicester City Transport operated 143 Dominators, the first of which arrived in 1977 and the last being withdrawn in 2005. London Buses operated a trio of Dominators as part of their Alternative Vehicle Evaluation trials; H1-3 (B101-3UVW) carried Northern Counties bodywork identical to those used by Greater Manchester, although no further orders resulted from the trial. As Daimler Fleetline production ended in 1980, several operators turned to the Dominator as a replacement. Central SMT took 51 Alexander-R bodied examples at East Kilbride and Old Kilpatrick depots, and Western Scottish (later renamed Clydeside) took 24 new at Greenock depot, before later purchasing a small number of vehicles second-hand from Central SMT. The Dennis Dominator was also exported, mostly to the Far East. China Motor Bus in Hong Kong used East Lancs and Alexander-RL bodied buses, whilst Kowloon Motor Bus used 40 Duple Metsec vehicles. Singapore Bus Services evaluated a 79-seat East Lancs example (SBS7003) in 1982, this later being shipped to Hong Kong and joining the Kowloon Motor Bus fleet in 1986.

The Dennis Dominator chassis provided the basis for three other Dennis products in the early 1980s. The Dennis Falcon was introduced in 1991, being unusually offered as a single deck bus ("Falcon H") and a double deck bus/coach option ("Falcon V"). Production of the Falcon lasted for just over two years, with the small number of UK operators including Ipswich Buses, Hartlepool Transport, Leicester Citybus,

London & Country and Midland Red North: 20 were also purchased by Kowloon Motor Bus in Hong Kong. The Dennis Dragon, also known as the Condor, was a three-axle double deck export chassis introduced in 1982 that was brought in great numbers by the Hong Kong operators China Motor Bus and Kowloon Motor Bus, and whose history is detailed separately within this publication. Finally, the Dennis Domino midibus was produced for less than a year in 1985/86 with only 34 being built in total. Greater Manchester PTE took 20 (1751-1770) with Northern Counties bodywork for 24 passengers, whilst South Yorkshire PTE received the other 14, with Optare bodywork seating 33 passengers.

Above: N715TPK was one of the last four Dennis Dominators, and was delivered to London & Country subsidiary Guildford & West Surrey in 1996. It carries East Lancs bodywork for 76 passengers.

After 1,007 Dominators had been built, of which only 37 were the single deck version, production ended in 1996 and turned instead to the short-lived Dennis Arrow, itself soon replaced by the Dennis Trident. The final Dominator was N716TPK in April 1996, the last of four East Lancs-bodied examples (DD13-DD16) for Guildford & West Surrey, this example subsequently worked for Arriva Southend and Arriva North West and Wales before finally being withdrawn from service in Merseyside.

Leyland Titan (B15) (1977-1984)

Leyland has started planning the replacement for the Bristol VR, Daimler Fleetline and Leyland Atlantean as early as 1973, even though all of these models remained in production for many years after that. The Leyland "B15" was designed as an integrally constructed double deck bus following a period of extended consultation with London Transport, who Leyland considered to be one of the most important customers for this new model. The first of five prototypes, NHG732P, was completed in 1975, and was fitted with a Park Royal dual door body with 71 seats. The remaining four B15 prototypes were also completed by Park Royal by 1977, with several different mechanical and bodywork specifications.

The "Leyland B15" was renamed the "Leyland Titan" at the start of production in 1977. The rear mounted Gardner 6LXB engine was positioned vertically with its radiator located above it on the offside, which resulted in a small, offset rear window that was a characteristic feature of the Titan's bodywork. Leyland's TL11 engine was offered as an alternative to the Gardner unit. It was originally intended that Park Royal would provide the bodywork, but slow production caused by labour disputes eventually led to Leyland having to expand their Leyland National factory at Workington in Cumbria and relocate Titan production there in 1980. Delays in the planned production schedule lead to a loss of some orders, as did London Transport's significant influence on the design which meant that potential operators were unable to choose their own bodywork supplier, and being built to a fixed height of 14'5" they were not able to specify a lowbridge option.

Above: London Transport T1101 (B101WUV) entered service from Bromley Garage in October 1984. One of the last Leyland Titans delivered to London, T1101 joined the Stagecoach Selkent fleet upon privatisation in 1994, and was finally withdrawn in 2001. After a period at Blue Triangle of Romford, which was subsequently acquired by the Go-Ahead Group, it was acquired for preservation in 2008.

Chassis numbers T1 to T282 (with a few numbering gaps) were bodied by Park Royal. 250 of these were for London Transport (fleet numbers T1-T250), and of the other remainder fifteen single door buses went to Greater Manchester PTE (ANE1/2/4/5T, FVR3V, GNF6-15V), five single door buses to West Midlands PTE (WDA1-5T) and two dual door buses to Reading Transport (YJB68/9T). A single Park Royal bodied Titan was completed for China Motor Bus of Hong King in mid-1980, being locally registered CD1213 and carrying fleet number TC1. Chassis number 300 onwards were almost exclusively for Leyland dual door bodied Titans for London Transport, although Reading ordered a further 10 (SBL70Y, RMO71-79Y) in 1983, and a single 1982 demonstrator (VAO488Y) was later sold to Ian Glass Coaches of Haddington (which was acquired by Lowland Scottish in 1991). The final Leyland Titan to be produced was B125WUV, London Transport's T1125, which was completed in November 1984.

The West Midlands PTE five were withdrawn in 1983 as being a non-standard type and returned to London where they became T1126-1130: in 1985 they visited Aldenham Works to be converted to coach seating for Selkent Travel's private hire and excursion fleet. Another second-hand acquisition was Leyland Titan demonstrator BCK706R, which joined Selkent Travel as T1131 in 1987 having been operated by John Fishwick & Sons of Leyland in Lancashire after its initial London trials completed in 1980.

Titan withdrawals commenced in 1992 as newer vehicles were delivered, and some were acquired by Home Counties based operators which had been successful in being awarded London bus contracts. Just two examples include London & Country who operated the 188 between Euston and Greenwich with Titans between 1992 and 1996, and London Suburban Buses who introduced second-hand Titans to route 4 between Archway and Waterloo until they acquired by MTL London in 1995. Away from London, the Titan found new owners with Merseybus, Oxford Bus Company and Swindon & District, amongst others. Other surplus Leyland Titans joined the fleets of London sightseeing tour operators such as Blue Triangle and the Big Bus Company. London's LEZ (Low Emission Zone) ended the option to use Titans in the Capital, although Big Bus Tours (as it is now named) were still using several Euro 3 compliant (after conversion) examples as static sales points around Central London.

MCW Metrobus (1977-1989)

Metro Cammell Weymann's Metrobus was a double deck bus of integral construction introduced to the UK market in 1977 being designed for heavy traffic routes within city areas. Built in Birmingham, the Metrobus was offered in four different lengths (9.7m, 11m, 11.3m and 12m), with most being powered by Gardner 6LX or Cummins engines connected to a Voith automatic transmission. The Metrobus also enjoyed some success as an export model, which resulted in some 11 and 12m versions being built as three axle "Super Metrobuses" for China Motor Bus and Kowloon Motor Bus in Hong Kong in the early 1980s.

In 1977, London Transport urgently needed a new double deck vehicle to replace older vehicles, and whilst the new Leyland Titan (B15) was preferred, the quantities available were not sufficient for this transition. MCW provided Metrobus demonstrator TOJ592S in December 1977 and being sufficiently impressed five Metrobuses to "London specification" arrived in 1978. Successful trials resulted in large orders on an annual basis between 1978 and 1985. Two batches were specified with luggage pens on the lower deck for the new Airbus services A1/A2/A3 to Heathrow Airport from 1980, and one batch were fitted with PA systems for use on tourist sightseeing services. London Transport trialled two Mk II Metrobuses in 1984 but did not progress to purchase any more. Instead, it purchased second-hand Mark I Metrobuses for tendered bus operations, these originating from Greater Manchester, Yorkshire Rider and Busways in Newcastle, and leased a further 29 for Harrow Buses services in 1987. Having operated nearly 1,500 examples, the Metrobus was retired from London service in 2004, providing many second-hand examples to bus operators throughout the country.

West Midlands PTE ordered over 1,100 MCW Metrobuses, including the first production bus (SDA831S) and five prototypes that were operated from Washwood Heath garage due to its proximity to MCW's factory. 50 of the fleet were dual-purpose Metrobuses with coach seats, and 14 others were initially delivered as guided buses for the Tracline 65 experiment. Between 1995 and 1999, over 600 Mk II Metrobuses were overhauled by Marshalls of Cambridge, extending their public service life until July 2010, when C903FON performed the final duty. Other UK Metrobus operators included Maidstone & District, Merseybus, Northern General, South Yorkshire PTE and Strathclyde Buses.

Above: London Transport was the largest operator of the MCW Metrobus, and M897 entered service in September 1983 from Shepherds Bush. In 1984 it passed to the privatised London General fleet and was purchased by Ensignbus for City Sightseeing work in 2000. Converted to open top format in 2002, it worked on Canterbury, South Coast and Windsor services, and is seen here passing Windsor Castle in 2007. It has since been sold to Red Bus in Cyprus.

China Motor Bus "Super Metrobuses" passed to New World First Bus in 1998, and immediately started to be withdrawn. These 12-metre-long dual door buses had seating for 108 passengers and being right-hand drive were soon purchased by overseas fleets. 2000 saw approximately half of their fleet travel to Australia for City Sightseeing tourist services, with many of the remainder soon heading to the UK for use by Big Bus Tours and Arriva London on their respective London sightseeing services. Both operators have only recently withdrawn their Super

Metrobuses, having replaced them with new purpose-built sightseeing buses.

MCW Metrobus production ended in 1989 with the sale of the company. The Metrobus design was subsequently purchased by DAF (chassis) and Optare (bodywork), who collaborated to redesign the model and create the Optare Spectra which was launched in 1991.

Leyland Olympian (B45) (1980-1993)

In 1979, Leyland was busy developing the successor to the Bristol VR (Bristol Commercial Vehicles became part of Leyland in 1965) based upon the Leyland Titan (B15) model that had been a significant element of the London Transport double deck fleet. This project was initially known as B45 but was soon re-named the "Olympian". It also addressed operator demands for a chassis-only product that could be completed with bodywork of their choice rather than being of integral construction.

The Olympian was offered in two standard lengths, 9.56m and 10.25m with highbridge (4.28m) and lowbridge (4.13m) options, and longer tri-axle versions were produced for export. Engine options were Leyland's 11.1-litre turbocharged unit, or Gardner's 6LXB and 6LXCT units. The chassis was similar to the Bristol VRT/SL3, although benefited from air suspension. ECW was the most common bodywork to be fitted, being the choice of larger fleets including London Buses, Lothian Regional Transport, Merseyside Transport, Strathclyde Transport and the Scottish Bus Group. Alexander received substantial orders for its R-type bodywork, mostly from Dublin Bus in Ireland as well as major Scottish fleets such as the Scottish Bus Group, Lothian and Strathclyde. Northern Counties bodywork was specified by Greater Manchester PTE, although their reliance on the type reduced as bus services were de-regulated, and Northern Counties found itself having to look for orders from elsewhere. East Lancashire Coachbuilders became the preferred option for many municipal fleets.

*Above: New to Northern General, C663LJR is an ECW bodied
Leyland Olympian which was part of a larger batch delivered in late
1985. It was subsequently purchased by Maghull Coaches of Bootle
in whose livery it is seen here in Liverpool City Centre.*

Leyland began bodying Olympians itself at Workington using ECW jigs
between 1988 and 1992, producing 197 vehicles looking very similar to
the original ECW design. Roe of Leeds, itself a subsidiary of Leyland,
also produced an Olympian body that closely resembled the ECW
design, and following Roe's closure and re-birth as Optare in 1984,
these continued in production until 1988 completing a combined total of
340 vehicles. Optare bodied Olympians were delivered to London
Cityrama, Maidstone Borough Transport, Reading Buses and Yorkshire
Rider. An unusual Leyland Olympian order was for 20 bodied by
Marshalls of Cambridge for Bournemouth Transport in 1981. In 1989,
Stagecoach took delivery of the unique "MegaDekka" Leyland
Olympian (F110NES), which had an Alexander body with 3+2 seating,

allowing for a seating capacity of 110 passengers (66 on the upper deck, 44 on the lower). It was originally allocated to Stagecoach Glasgow for school duties, was transferred to Stagecoach East Midland in 1990, and them Bedford in 1993, where it remains in service at the time of writing (2012).

Above: Leyland Olympian GKE442Y was part of a batch of ECW-bodied double deck coaches delivered to Maidstone & District between 1983 and 1986 for use on Invictaway commuter services between the Medway Towns and London. Upon withdrawal many were sold to Clydeside Scottish, including this example.

ECW also developed a coach option on the longer Olympian chassis with dual-purpose coach seating. Demonstrator ADD50Y was provided to National Bus Company subsidiary Wessex in 1982, who subsequently loaned it to other NBC operators for evaluation on various National Express coach services. Ultimately it was not selected for

National Express duties, and the decision was made to purchase double deck MCW Metroliner coaches instead. The Olympian coach did find favour with many operators of London commuter services, including Alder Valley (Londonlink services), Eastern National (including London Express services), London Country (Green Line services 711 to Harlow, 720 to Gravesend and Luton Flightline 757) and Maidstone & District (Invictaway services to London). London Coaches purchased two coaches in 1986, although these were both sold to Southampton City Transport for their Red Ensign fleet in 1990.

Leyland also produced a tri-axle Olympian for export orders in three different lengths of 10.4, 11.3 and 11.9 metres, adding an air-conditioned version in 1988 which made it more suitable for operation in hot and humid climates. A number of Alexander RH bodied Olympians were operated by Kowloon Motor Bus and Citybus in Hong Kong until the early 2000s, and upon withdrawal many were shipped back to the UK as their right-hand drive format made them suitable for UK roads. Their length and high seating capacity (100+ passengers) made them attractive to operators with busy routes, and Stagecoach used them to help establish their "Megabus" brand of low-cost, long distance coach services, as well as on budget low cost bus services such as "Magic Bus" in cities such as Manchester. Other operators took advantage of their capacity by using these for school transport and rail replacement services, indeed Stagecoach Hampshire applied South West Trains livery to some for rail replacement duties for their own SWT rail franchise. London's Big Bus Company purchased 21 in 2005/6 for use as open top sightseeing buses in the Capital.

Leyland Olympian production ended in 1988, when Leyland was purchased by Volvo. For a while, Volvo carried on building the Olympian at Workington to complete outstanding orders. The Olympian name continued with the new Volvo Olympian which was built in Irvine in Scotland from 1993. Whilst the Leyland examples have all but disappeared from service, the last few years have seen many former Dublin Bus Alexander bodied Volvo Olympians being brought to the UK under disposal agreements with Ensignbus of Purfleet and Southdown PSV of Copthorne, resulting in hundreds being sold to a wide variety of smaller UK operators.

Leyland Tiger (B43) (1981-1993)

Leyland developed the Tiger chassis for two reasons, to replace the ageing Leyland Leopard which has been in production for more than 20 years, and to provide a realistic competitor to Volvo's B58 which was gaining popularity amongst coach operators seeking more performance at the start of the 1980s. The Tiger was essentially a heavily re-worked Leopard, offered with a turbocharged Leyland TL11 engine and hydracyclic semi or fully automatic gearbox, air suspension and a re-designed driver's area. At launch the TL11 provided only 220hp, rising to 245hp soon after and then 260hp in 1984. Also in 1984 came the option of a Gardner 6HLX engine, and in 1987 the North American Cummins L10 engine with manual or automatic ZF gearbox finally gave the Leyland Tiger a decent performance rating of 290hp. Volvo acquired Leyland in 1988, and within a year the Volvo THD100 engine was added as an option, at the same time the Leyland and Gardner units were dropped.

The coach version of the Tiger was a popular choice for hundreds of different operators, and for much of the 1980s, it was the backbone of the National Express long-distance coach network, with smart Plaxton and Duple bodied examples providing standard and Rapide services throughout the country. One of the single largest Tiger coach fleets was created by London Country's Green Line, which took large batches with different bodywork, including ECW's B51, Plaxton's Paramount 3200 (short and long) and Paramount 3500, Duple's Dominant and 320 and Berkhof's high-floor Everest. Upon initial withdrawal, many of London Country's ECW and Berkhof examples found themselves being dispatched to fellow Drawlane Group company East Lancashire for rebodying. Emerging with new bus bodies seating between 49 and 61 passengers, they stayed within the Drawlane Group passing to Midland Red North.

Above: WPH139Y was new to London Country as a Green Line coach in 1982, carrying ECW's B51 bodywork with 53 coach seats. It passed to London Country South West at Dorking upon privatisation in 1986 and was sold to Midland Red North in 1989. It was rebodied by East Lancs with a new bus body in 1990 and served with Arriva Midlands North before being sold on several times more.

The bus version of the Tiger had a less powerful engine than its coach counterparts, and conventional leaf springs replaced air suspension. The most significant customers of the bus version were Citybus and Ulsterbus in Northern Ireland, who were looking for a replacement for the retired Bristol RE. From1984 these were bodied locally by Alexander (Belfast) with N-type bodywork, and from 1990 to 1993 with the newer Q-type body. Other operators specified bus bodywork from Alexander, Duple or Plaxton, and East Lancs offered re-bodied buses to older Tiger chassis that had previously carried coach bodywork. The

Ministry of Defence was a significant user of the Tiger, having a fleet of Marshall, Plaxton and Wadham Stringer bodied examples.

When Volvo purchased Leyland in 1988, it brought the Leyland Tiger and the Volvo B10M together. With only 3,500 Tigers having been sold against over 20,000 B10Ms in the same period, Volvo decided to end Leyland Tiger production in January 1993.

Dennis Dragon / Condor (1982-1999)

Developed as an export model based upon the existing Dennis Dominator chassis, the Dennis Dragon was a three-axle double deck bus chassis designed for the Far East markets. The Dragon was originally intended for Kowloon Motor Bus, but with Kowloon in Chinese meaning "nine dragons", China Motor Bus insisted upon it being renamed to the "Dennis Condor" to avoid them having to operate a bus named after one of their competitors. Powered by either a Gardner or Cummins engine connected to either a Voith or ZF transmission, early models had two additional rear wheels fitted ahead of the rear axle, but after 1986 a complete additional axle was installed instead. Five prototypes were completed in 1982, three Dragons with incredible Alexander 128-seat bodywork for Kowloon Motor Bus (3N1-3N3) and two 108-seat Condors, one bodied by Duple Metsec and the other by Alexander, for China Motor Bus (DL1 and DL2). Aside from the four Alexander bodied demonstrators, all other Dragons and Condors were fitted with Duple Metsec bodywork, supplied in kit form and built locally by the operator, many being fitted with three sets of doors.

Over the following 17 years, Kowloon Motor Bus went on to acquire a total of 765 Dennis Dragons, all with Duple Metsec bodywork, in three different lengths of 9.9 metres (ADS class, also known as "Baby Dragons"), 11 metres (S3N class) and 12 metres (3N class). China Motor Bus brought less, with 166 Dennis Condors of either 11 (DM class) or 12 metre (DL class) length, and all of theirs passed to New World First Bus after China Motor Bus lost its operating franchise in 1998. Citybus ordered the Dennis Dragon in the early 1990s, with 40 short 10.3-metre examples and 80 12 metre buses entering service between 1994 and 1998. Four 12 metres buses were operated by Hong

Kong Air Cargo Terminals for staff transport from 1996, and when this service ended in 1999 all four passed to New World First Bus for further use.

Above: London's Big Bus Tours acquired a number of Duple Metsec bodied Dennis Condors that were new to China Motor Bus in Hong Kong. These were used on London sightseeing services, where their large passenger carrying capacity was put to good use, as seen here with fully loaded G954FVX (former China Motor Bus DL37).

Over the following 17 years, Kowloon Motor Bus went on to acquire a total of 765 Dennis Dragons, all with Duple Metsec bodywork, in three different lengths of 9.9 metres (ADS class, also known as "Baby Dragons"), 11 metres (S3N class) and 12 metres (3N class). China Motor Bus brought less, with 166 Dennis Condors of either 11 (DM class) or 12 metre (DL class) length, and all of theirs passed to New World First Bus after China Motor Bus lost its operating franchise in

1998. Citybus ordered the Dennis Dragon in the early 1990s, with 40 short 10.3-metre examples and 80 12 metre buses entering service between 1994 and 1998. Four 12 metres buses were operated by Hong Kong Air Cargo Terminals for staff transport from 1996, and when this service ended in 1999 all four passed to New World First Bus for further use.

The Dennis Dragon was also selected by Stagecoach for operations in two African countries. Ten Duple Metsec bodied Dragons with 115 seats were shipped to Stagecoach Malawi in late 1992, taking fleet numbers 2001-2010. A further 20 joined the Stagecoach Kenya fleet in 1995/96, these also being 115-seat examples completed by AVA (Association of Vehicle Assemblers) using Duple Metsec kits and being allocated fleet numbers 201-220.

The Dennis Dragon/Condor was superceded by the new three axle Dennis Trident 3 in 1997, although the final examples were completed for Kowloon Motor Bus in July 1999. As their original operators started to withdraw them, with non-air-conditioned examples generally being retired first, their young age and right-hand drive format made them a suitable vehicle for the UK market. All of Stagecoach Kenya's Dragons were shipped to the UK between 1998 and 2000 to join Stagecoach's low cost "Magic Bus" operations in Manchester and Glasgow, the last of these remaining in service in Manchester until 2010. Weavaway Travel of Newbury in Berkshire purchased seven former China Motor Bus 106-seat Dennis Condors for their "Noah Vale" school bus fleet, retiring them in 2006 before being selling them to other UK operators. Big Bus Tours acquired a number of former China Motor Bus Dennis Condors, still having no less than 16 of them in use on London sightseeing services at the time of writing, their huge seating capacity for 107 passengers (72 upstairs, 35 downstairs) making them ideal for transporting large numbers of tourists.

One former China Motor Bus/New World First Bus Dennis Condor has been preserved since 2003, DM17 (indicating a "medium" 11 metre length chassis) was registered ES997 in Hong Kong and was acquired upon withdrawal to become H74ANG in the UK.

Quest 80 (1982-1985)

With its origins as a Telford-based design and engineering consultancy, Quest 80 initially experimented with bus, coach and trolleybus chassis designs for use in South Africa, later moving onto products for the UK domestic market. With Ford exiting the bus industry, Quest 80 attempted to fill their void, using a wide variety of Ford components in their designs. Perhaps the most bizarre aspect of Quest 80 designs was the "U-Drive" configuration, with the engine and gearbox being mounted in opposite rear corners, with a chain drive connecting the two. Quest 80 production included:

- B: low-floor midibus, Perkins-engined, typically B23F
- C: coach format, Ford 360-engined, available as 9, 11 or 12 metres long
- D: midibus format, Ford 360-engined, typically B33F
- J: midi-coach format, Ford Cargo-engined, Jonckheere Piccolo C37F body
- VM: coach format, Ford 360-engined, Plaxton Paramount 3200 C53F body

There are only two UK-based operators of Quest 80 products to be noted. In the first half of 1984, Excelsior Coaches of Bournemouth took delivery of seventeen 12-metres VM coaches (of a batch of twenty ordered), being A807-823LEL, although A823LEL operated for just over a month before being destroyed by fire. Despite having Ford Sabre marine engines and conventional driveshafts in place of the chain-drive arrangement, the batch are recorded as having significant and ongoing reliability issues, with Excelsior later acquiring several more as a source of spares to keep the fleet operating. Merseyside PTE acquired six B-specification midibuses at the turn of 1984-5, with Locomotors B23F bus bodywork. B927-932KVM (932 later being re-registered to C844OBG) were acquired for disabled transport but were later sold on having not seen much use in service. 927 and 931 later worked for Millers of Cambridge, 928 with for Somerbus of Paulton and 930 for Luckettts (Luckybus) of Watford.

Above: A813LEL is the sole-surviving Quest 80 VM, with Plaxton Paramount 3200 coach bodywork. One of the 1984 delivery for Excelsior Coaches of Bournemouth, it is now preserved.

Quest 80 Limited ceased trading in 1985, but many of their chassis were bodied later and could be found with registration dates between 1986 and as last as 2001, especially for some of the J-specification coaches that ended up in Cyprus. Two Quest 80 survivors are preserved: A813LEL from Excelsior Coaches is the sole surviving VM coach, whilst C844OBG is one of the Merseyside PTE examples.

Leyland Lynx (1984-1992)

Leyland developed the Lynx single deck bus as a replacement for the ageing Leyland National. Whereas the National was of integral construction, the Lynx had a separate chassis and was able to be finished by different bodybuilders for export orders, although UK operators were provided with Leyland's standard bodywork noticeable by its raked driver's windscreen reminiscent of buses in the 1950s. The Lynx was 12 metres long with a horizontally mounted rear engine, provided by either Cummins, Gardner or Leyland. The radiator, however, was mounted at the front of the chassis to provide the most effective cooling, and on the updated Leyland Lynx II, launched in June 1990, the radiator grille was located behind a protruding front panel. The redesigned Lynx II also offered a Volvo engine option, which offered lower emissions but also a reduction in performance.

Above: Halton Transport of Cheshire is one of the UK's last municipal bus operators. Halton had been a major operator of the Leyland Lynx, having bought 36 between 1986 and 1992, including K853MTJ, the last Lynx to be built. Preserved example H35HBG is a Mark 2 Lynx, distinguishable from earlier models by the protruding front grille

Over 900 Leyland Lynx I buses were built between 1984 and 1990, and other than West Midlands PTE, with over 250 examples placed into service, no single operator built up a significant fleet of the type. Lynx I orders were received from over 30 different operators, including many municipal concerns. Whilst never a common vehicle in the London Buses fleet, the de-regulated bus industry saw many Lynx being placed into service on tendered services with Home Counties operators such as Boro'line Maidstone, Grey Green, Jubilee Buses, Kentish Bus, London Buslines, London & Country, Luton & District and Thamesway (part of Eastern National). The Lynx II was less successful, with only 140 orders being received from a much smaller number of operators.

A small number of Lynx I received non-Leyland bodywork. Citybus in Belfast ordered six Alexander (Belfast) N-type examples, which were registered HXI3007-12, and Singapore Bus Service ordered a single Lynx with Alexander PS-type 53-seat dual-door bodywork in 1989 registered SBS3572Y. UTC in Sydney also ordered a Lynx for trials, having PMC Metro 90 dual-door bodywork, although delays in production lead to MO7936 never entering service and it was sold to Premier of Illawarra. Australia also became the home for two new Leyland-bodied Lynx in April 1990, being registered ZIB730/1 with Action of Canberra.

Having been succeeded by the Volvo B10B, Lynx II production ended in 1992 with the last example joining Halton Transport as K853MTJ. The remains of the once large West Midlands fleet were withdrawn in early 2009, although ten were temporarily retained as driver trainers for a further year. At least four Lynx II are known to remain in active service into 2012: J268UDW (new to Cardiff) with Nu-Venture of Aylesford in Kent and H48NDU, J724KBC and J295TWK (all acquired from Hedingham Omnibus) with Regal Busways of Chelmsford in Essex. A small number of Leyland Lynx have been saved and restored by preservationists.

MCW Metrorider (1986-1989)

Metro Cammell Weymann of Birmingham launched their Metrorider midibus design at the 1986 Motor Show. It was an integrally constructed bus available in two lengths of 7.0 and 8.4 metres being powered by either a Perkins engine with a ZF manual gearbox or the more powerful 5.9-litre Cummins B-Series engine with an Allison automatic transmission. The Metrorider design was notably different to other similar sized buses of the time which visibly owed their origins to commercial van conversions: in common with the earlier Optare CityPacer it incorporated an angled front with large windscreen, large side windows providing a bright interior, and clean bodywork lines.

The first two Metroriders were produced in October 1986 as demonstrators (D482/483NOX) and significant orders followed from Greater Manchester Buses, Strathclyde Buses and West Midlands Travel, as well as many municipal fleets and recently privatised former National Bus Company operators. London Buses ordered 134 Metroriders in a mixture of both lengths, the first 22 of which (MR1-22) started work for Westlink on "Kingston Hoppa" local services K1-K3 in June 1987. Subsequent batches became Harrow Hoppas, Bexley Hoppas and Walthamstow Hoppas: others were allocated to Orpington's Roundabout services and a Wandsworth Health Authority contract for six vehicles to serve St George's Hospital in Tooting. MCW Metroriders became a common choice of vehicles for Home Counties operators of deregulated services, including Kentish Bus (using Metroriders borrowed from Northumbria), Londonlinks (including 11 Metroriders purchased from Darlington Corporation Transport) and London Country North West, who operated 39 of the type mostly from their Garston and Hemel Hempstead garages. Two examples were sent to Hong Kong in 1988, DY6050 becoming Kowloon Motor Bus AMR1 and DY835 joining China Motor Bus as CM1. The last Metrorider produced before production was taken over by Optare was F171DET, delivered to BTS Borehamwood in May 1989.

Above: MCW built the Metrorider midibus between 1986 and the closure of MCW in 1989, with the design being acquired by Optare. Optare continued to build the MetroRider until 2000. W675DDN, new to Jones of Pontypridd in 2000, was one of the last MetroRiders to be built, later joining Midland Classic of Burton upon Trent.

The Metrorider was a popular UK midibus, and was produced until MCW was closed in 1989, with the designs then being purchased by Optare. The new owners continued to produce the Optare MetroRider (with a middle upper case "R") until 2000 by which time their new Optare Solo had become established as an alternative.

CVE Omni (1988-1999)

The Omni was a low-floor minibus built by City Vehicle Engineering (CVE) of Shildon in County Durham. It had an integral design to achieve a low floor height – the 6.6-metre-long version was just under 30cm from the ground – assisted by being front wheel drive and running on 16" wheels. It was powered by a four-cylinder 2.5-litre Land Rover engine linked to a ZF four-speed gearbox. Primarily intended as a welfare vehicle, the front suspension of coil springs was a contrast to rear air bag suspension, which allowed for the rear of the minibus to be lowered so that wheelchair users could board via a ramp instead of a wheelchair lift. The Omni was typically configured with 16 seats in welfare configuration, or 23 for commercial operations – the latter being easy to identify from their over-sized front destination display box. The Omni design was a result of CVE purchasing existing design rights from Austrian manufacturer Steyr.

Above: H389KPY is a welfare-specification CVE Omni, with DP17FL configuration including a wheelchair lift at the rear. It was new to Empress Coaches of St Leonards (near Hastings) in September 1990.

Amongst the initial 25 ordered, Durham County Council took three, including two for local bus services funded by the Rural Development Commission. In March 1989, the London Borough of Hounslow purchased three for a council-funded initiative to provide London's first ever low-floor, wheelchair-accessible bus route, which was operated by Westlink as route H20. Another trio were funded by the London Borough of Richmond for new routes R61 and R62. Unfortunately, the CVE Omni was not known for its reliability in service in the Capital, and London United (who had since acquired Westlink and the H20) replaced them with space MCW Metroriders. CMT of Aintree were the largest PSV operator, using six for services in Liverpool.

Whilst popular in the welfare market, a lack of order from commercial operators means that CVE went into liquidation late in 1990. By the middle of 1991, however, the operation has been taken over by new company "Omni Bus", which completed sixteen partly built Omnis purchased from the liquidator, all to welfare specification. Omni Bus improved the existing heating and ventilation systems, and a side-kneeling version based on the original 6.6-metre Omni was also developed, with Kent County Council showing early interest. In 1992 a new longer (7.7-metre) six-wheeled version was added, which allowed for seating capacity to be increased from 23 to 29 with an additional 11 standing (total capacity of 40). Two (M150EAV and M589SDC) were supplied to Whippet Coaches of Fenstanton in January 1995, and a trio of electric Omnis (fitted with battery packs) joined the Strathclyde Fleet as EV1-EV3 (R967JGA, R571JGG, R570JGG) in the first half of 1998.

The new owner focused more on the welfare variant, with sizeable orders coming from Dorset, Humberside, Kent, Leeds, Leicestershire, Newcastle, Northamptonshire and Sandwell councils. However, a decline in sales blocked any further investment, and routine production slowed in 1996. Numerically the last Omni chassis (454), S922JSH was the final B16F bus delivered, this one to the Scottish Borders Council as a welfare vehicle in February 1999, with the company ceasing trading soon after. Instead, the task of bringing low-floor minibuses to the market was progressing at speed by the rival Optare Solo ...

Dennis Dart & Dart SLF (1989-2008)

The Hestair Group, owners of both Dennis and Duple, planned to build a new "midibus" at the end of the 1980s, hoping to fill the void between the minibuses such as the Ford Transit and full- sized single deck buses. The Dennis Dart was launched in 1989, fitted with a Cummins 6BT engine and an Allison gearbox, and available in two lengths of 8.5m and 9.8m. The bodywork was a new design, the Duple Dartline, which was destined for a very short life as Duple was purchased by Plaxton later the same year. Plaxton did not want to retain the Duple Dartline rights, which instead passed to Carlyle of Birmingham in 1991, who continued producing their own style of bodywork until production passed to Marshall of Cambridge in 1992.

The Dart was also available with other bodywork, including Wadham Stringer's Portsdown, Wright's Handbybus, Northern Counties' Paladin, Alexander's Dash and East Lancs' EL2000. Reeve Burgess produced their Pointer bodywork, which was renamed to "Plaxton Pointer" when production switched to Plaxton's Scarborough premises in 1991. Many bodybuilders offered dual door options. Demand for the now ubiquitous Dennis Dart grew steadily until the mid-1990s, when the need for newer low-floor buses with increased accessibility saw a decline in orders. Production of the original step entrance Dennis Dart finally ended in 1998, as the newer Dennis Dart SLF (Super Low Floor), launched in 1995, had effectively superceded it.

The Dart SLF retained the original Dart mechanical systems, apart from a new air suspension system, and was available in three different lengths (9.2m, 10m and 10.6m). It was fitted with a low-floor Plaxton Pointer body, which was upgraded to Pointer 2 in 1997. As with the original Dart, low-floor bodywork was provided by different companies, including the East Lancs Spryte, Caetano's Compass and Nimbus, Wright's Crusader, Alexander's ALX200 and Marshall's Capital. 1997 also marked the introduction of the longer Dart SPD (Super Pointer Dart) at 11.3m long, and the shorter MPD (Mini Pointer Dart) at 8.8m long. From 2002, a narrower Dart SLF with modified East Lancs or Caetano bodywork was produced to meet the 7'6" width restrictions on the Channel Islands of Guernsey and Jersey, the Caetano version also being used by the Gibraltar Bus Company for similar reasons.

Above: London Buses ordered 153 Dennis Darts with Reeve Burgess bodywork in 1991/1992, being 8.5-metres long and 2.25 metres wide. J136DUV was new in June 1992 and was privatised to London United in 1994. Upon withdrawal in 2001 it was purchased by the Anglia Bus Company and changed owners again in 2004 to Sunray Travel of Worcester Park who used it on rural services in Surrey.

With a tightening of emissions legislation in 2001, Euro 3 compliant 3.9-litre four-cylinder Cummins engines were introduced on all Dart SLFs except the SPD which was given a more powerful 5.9-litre six-cylinder version. Cummins Euro-4 compliant engines were introduced in 2006. Dennis Dart SLF production ended in 2008, itself having been superceded by the Alexander Dennis Enviro 200 Dart which was launched in 2007. The last Dennis Dart SLF produced was LU2001 for Park Island Transport in Hong Kong.

Above: London General LX05EYS is a low floor Dennis Dart SLF with Alexander Pointer bodywork, new in April 2005. LDP265 was allocated to Waterside Way Garage and is seen in this view in Clapham, South London operating route G1 to Streatham.

Optare Solo (1997-2012)

Optare of Leeds launched their low floor Solo midibus in 1997, the name reflecting its low floor level, at only 20cm with kneeling suspension. The Solo is an integral bus, powered by a Cummins 5.9-litre, MAN 4.6-litre or most commonly a Mercedes 4.25-litre engine which is coupled to an Allison automatic gearbox. The Solo was produced in eight different lengths:

Optare Solo Specifications

- M710, 7.1 metres, 23 passengers ("SE", introduced in June 2006)
- M780, 7.8 metres, 25 passengers
- M810, 8.1 metres, with various seating capacities
- M850, 8.5 metres, 29 passengers
- M880, 8.8 metres, with various seating capacities
- M920, 9.2 metres, 33 passengers
- M990, 9.9 metres, 37 passengers
- M1020, 10.2 metres, 37 passengers (additional 30cm needed to accommodate Cummins engines)

Above: Thames Travel of Wallingford operate several Optare Solos within their fleet. YJ10 MFE is a Solo M880 from 2010. The type was operated on many of Thames Travel's services, although the 128 between Reading and Wokingham was one of the most common.

Following on from two development chassis, Optare built two demonstrators, one M850 (S794XUG) and one M920 (S903DUB), and the first order for 32 M850s came from Wilts & Dorset in May 1988 (2601-32, R601-22NFX and S623-32JRU).

Standard Solos are 2.5 metres wide, although a narrower "Slimline" version at 2.33 metres wide was introduced in 2004 which can be identified by "SL" being added to the vehicle's code, demonstrator YN04XZB being the first. Optare introduced the 7.1m shorter "Solo SE" in June 2006. The Solo SR was added in October 2007, which provided a restyled front-end design with similar lines to the Optare Versa, and in November 2008 the new "Solo +" was unveiled at the Euro Bus Expo at the NEC in Birmingham. Whilst the Solo SR continued to sell well, the Solo + was not well received and did not progress beyond the single prototype model.

A hybrid option was introduced in 2005, with electric propulsion provided by Eneco/Traction Technology and supported by a small diesel engine. Demonstrator GX55LNF carried silver "Traction Technology" livery and was used to increase the frequency on Arriva Southern Counties' Horsham Park & Ride service (98) in West Sussex from late 2007. Traction Technology went into administration after a few months, and the vehicle was seized as an asset and sent for disposal: by the end of 2009 it was with Compass Royston of Stockton, where it was converted back from hybrid to conventional diesel power.

The next environmentally friendly Solo was the all-electric Solo EV, introduced in 2009, which used an Enova Systems P120AC induction motor powered by two banks of lithium-ion phosphate batteries. Demonstrator YJ09EZR was evaluated by several operators around the country including Nottingham City Transport and Johnsons of Henley-in-Arden. Under the Government's Green Bus Fund, the Solo EV has since been ordered by Nottingham City Council, Durham City Council (for "Cathedral Bus" services), Travel de Courcey of Coventry and Dorset County Council.

Above: With many operators now seeking more environmentally friendly buses, Optare developed an electric version of its popular Solo midibus in 2009. YJ09EZR was subject to an early trial which took place with Johnsons Coach & Bus Tours in Stratford-upon-Avon, which involved Aston University (assessment of practical results such as emissions and noise pollution) and Coventry University (human aspects, including the reactions of passengers, pedestrians and other road users). YJ09EZR was subsequently evaluated by other operators.

As a consequence of working capital challenges, Optare agreed to sell nearly half of its shareholding to the Indian Hinduja brothers, (owners of the Indian Ashok Leyland Company) in January 2012, taking their shareholding in Optare up to 75%. The Optare Solo has since been made available to the Indian market, launched as the "Ashok Leyland Solo" at the 2012 Delhi Auto Expo. Powered by Ashok Leyland's H-Series engine, these Solos are notable for carrying the Leyland "wheel" badge on their front panel.

January 2012 saw Optare announcing the end of production of the Optare Solo after 14 years, pending the imminent launch of the new "Solo SR" range later that year. The "new look" Solo SR has been provided to both UK and export customers, is available in Slimline format in four different lengths (7.1, 7.8. 8.9 and 9.6 metres), with the longer two chassis also available as a wider 2500mm option. Mercedes and Cummins engines continue to be offered, as will Allison transmissions.

Dennis Trident 2 & 3 (1999-2006)

In 1997, Dennis unveiled a new low floor double deck bus chassis to replace the short-lived Dennis Arrow. Known as the "Dennis Trident", it was fitted with a Cummins Euro 2 or Euro 3 compliant engine, transversely mounted to the right of the engine bay, and drive was provided by a Voith or ZF transmission. Available in four lengths of 9.9, 10.5, 10.6 or 11.4 metres, the Trident was most commonly completed by Plaxton with their "President" bodywork or Alexander with their "ALX400" body, both being available as single or dual-door format. East Lancs and Northern Counties also bodied a small number. When Dennis was acquired by TransBus in 2001 the chassis became known as the "TransBus Trident" and the Alexander bodywork simply as "TransBus": a further change of control following TransBus International's collapse in 2004 lead to the "Alexander Dennis Trident".

Throughout the first years of the new century, the Dennis Trident became the most popular choice of double deck bus chassis for the UK's major bus fleets, with significant orders from Arriva, First Group, Go Ahead, Lothian Buses, Stagecoach and Travel West Midlands, amongst many others. The very first examples were delivered to East London's Leyton Garage in January 1999, these having Alexander 73-seat dual door bodywork and starting the "TA" class. Stagecoach remained the largest purchaser of the Trident to the end of its production, by which time most other fleets had sought alternative new buses.

Above: The Stagecoach Group acquired the Tyne & Wear operator Busways in July 1994. V157DFT, an Alexander ALX400 bodied Dennis Trident, was delivered in 1999, and is typical of the standard Stagecoach double-deck bus specification of the time. 17657 later passed to Stagecoach Manchester, where it worked on the low-cost Magic Bus network of services and is seen here in Piccadilly Gardens.

Unusual UK Tridents included L1TSB, built new as an exhibition vehicle for Lloyds TSB with East Lancs bodywork in 2000, PF04WMK, which was an East Lancs bodied "Crechemobile" used in the Gateshead area from 2004 and BX05CGG, a new Handsworth Playbus in 2005, bodied by Alexander. Outside of the UK, Dublin Bus ordered ten Tridents with TransBus ALX400 dual door bodywork in 2003: DT1-10 (03-D-10001 etc) were considered an unusual choice as the first Dennis chassis in their fleet. Several East Lancs open top Dennis Tridents were purchased for sightseeing duties in Spain by Barcelona Tours and Madrid Vision.

The UK version of the Dennis Trident was also known as the "Trident 2" indicating it had two axles. This differentiated it from the three-axle "Trident 3", an export chassis produced to replace the Dennis Condor/Dragon, which was Dennis's previous export chassis until 1999. The Trident 3 was offered in four different lengths of 10.3, 10.6, 11.3 and 12.0 metres, with the additional rear axle added to reduce the rear overhang and address the increased weight of the larger bus. The Trident 3 was built with bodywork either by Duple Metsec (DM5000) or Alexander (ALX500). Six Trident 3 prototypes were completed in 1996, with three being sent to Citybus in Hong Kong and two to Kowloon Motor Bus.

Above: Y858GCD is a 2001 Dennis Trident with Plaxton President bodywork. Numbered 859 in the Brighton & Hove fleet, it was named "Thomas Tilling" to remember the founder of the Tilling Company which commenced operations in Brighton in 1915. It is seen in this view collecting passengers in North Street, Brighton in 2011.

Kowloon Motor Bus went on to purchase the largest fleet of Dennis Trident 3s, closely followed by New World First Bus (formerly China Motor Bus) and then Citybus who took a much smaller number. 150 of Kowloon's initial delivery of 336 Tridents were allocated to their Long Win Bus fleet for a network of services to the new Hong Kong International Airport. In other parts of the world, BC Transit of British Columbia used 29 Duple Metsec bodied Trident 3s from Victoria and Kelowna, and Singapore Bus Services took delivery of 20 Duple Metsec examples for use in the east of the country. US sightseeing operator Gray Line ordered 40 Duple Metsec for sightseeing services in San Diego and New York, with all examples later moving to New York.

The Trident 2 chassis was subsequently redeveloped into a new chassis by Alexander Dennis in 2005, which was marketed as the "Enviro 400", although also still being known as the Trident 2 (the details of which are documented separately in this publication). The last "Pre-Enviro" Tridents were delivered to Isle of Man Transport in 2006, HMN244-249J having East Lancs single door bodywork. By this time, Trident 3 chassis had already been retired from production and had been replaced by the integrally constructed TransBus (and later Alexander Dennis) Enviro 500. The last Trident 3 was completed in April 2003 for Kowloon Motor Bus, KX2356 having fleet number ATR392.

Alexander Dennis Enviro 400/400H Trident (2005-Current)

The Alexander Dennis Enviro 400 Trident is a low-floor chassis and body combination built by Alexander Dennis, which is also available as a separate chassis (known as the "Dennis Trident 2") for receiving bodywork from other manufacturers or as a double deck body for fitting onto other manufacturers' chassis including Volvo (B7TL/B9TL) and Scania (N230UD). The Enviro 400 Trident was originally designed by TransBus before it closed in 2004 and was subsequently reformed as Alexander Dennis, and the Enviro 400 bodywork was introduced to be a replacement for the existing Alexander ALX400 and Plaxton President ranges. Whilst the Alexander Dennis chassis on its own is known as the

Dennis Trident 2, the complete vehicle is simply referred to as the "Enviro 400" within Alexander Dennis promotional literature.

Above: The Stagecoach Group have been the largest purchaser of Alexander Dennis Trident 2/Enviro 400 and 400H hybrids. The latter are allocated to busy city centre services, and YN61BGF was a 2011 arrival for Stagecoach in Sheffield, being based at the Holbrook Garage. It carries a modified livery for "a cleaner greener Sheffield".

Alongside other members of the Alexander Dennis range, the chassis and mechanical units are constructed at their factory in Guildford, Surrey and then transported by road to their Falkirk premises for the addition of bodywork. The Enviro 400 Trident is available in three lengths of 10.2, 10.9 or 11.4 metres, and is 2.55 metres wide. Alexander Dennis versions are powered as standard by a Cummins six-cylinder 6.7-litre EEV engine attached to either a Voith four-speed or ZF six speed automatic gearbox. The Enviro 400 bodywork has been designed to maximise seating capacity, which varies between 67 and 79 passengers, dependent on individual layout, and is available in single and dual door format. The aluminium body panels and glazing are bonded, and the front and rear panels are both GRP mouldings.

Above: Alexander Dennis constructed this distinctively liveried Enviro 400H Hybrid demonstrator in October 2009: it is dual- door and seats 37 passengers on the upper deck and 27 on the lower.

A hybrid electric version was also made available, the Enviro 400H. Following collaboration between Alexander Dennis and BAE Systems, this was designed to include a 4.5-litre Cummins EEV engine and a BAE Systems HybriDrive driveline capable of a peak power output of 195kW. The hybrid model reduces both fuel consumption and greenhouse gas emissions by approximately one third over a conventional diesel engine. The 400H's lithium ion batteries do not require mains charging during their normal service life, with the bus regenerative braking system channelling energy back into them from normal driving activities. Whilst the hybrid model is more expensive than the standard diesel-powered bus, many operators have successfully taken advantage of the Government's Green Bus Fund scheme which contributes towards the price difference.

One notable early Enviro 400 Trident was LX55HGC, one of the very first models built in 2005, which was used by Stagecoach East London to replace the bus destroyed in the 7th July 2005 terrorist bombings in London, and which entered service from Stratford Garage in January 2006 named "Spirit of London". Metroline were the first fleet to take delivery of a completed Enviro 400 Trident, also in January 2006, with LK55 KJV receiving fleet number TE665.

A single Enviro 400 Trident was to be involved in a special event in 2008, an overland expedition to China to attend the closing ceremony of the 2008 Beijing Olympic Games and celebrate the countdown to the 2012 Olympic Games in London. Of 1,000 bus driver applicants to drive the journey, 12 were selected for the route which would travel across mainland Europe into Turkey, then through Georgia, Azerbaijan, Turkmenistan, Uzbekistan, Kazakhstan and Kyrgyzstan before reaching China. However, the estimated £450,000 cost of the expedition was considered excessive, and the newly elected Mayor of London Boris Johnson cancelled the project. Instead, Alexander Dennis constructed a special Enviro 400 shell which was then sent to a specialist conversion company in Leeds before being shipped to China. It was watched by millions of people around the world opening up to showcase the London skyline and landmarks, the voice of Leona Lewis, the guitarist Jimmy Page and the football talents of David Beckham.

The Enviro 400 and 400H Tridents have become the standard double deck bus within London, with significant fleets currently being operated by Abellio London, Arriva London, First Group, Go-Ahead London, Metroline, Quality Line, RATP London United and Stagecoach London. The 1000th Enviro 400 Trident for London was displayed at the 2011 Showbus event, LX11BKN joining Stagecoach London as 19819. Away from the Capital, the model's appeal has seen it introduced into the fleets of national operators Arriva, First Group, Go Ahead and Stagecoach, other significant operators (such as East Yorkshire, Lothian Buses, National Express West Midlands and Reading Buses) as well as smaller independent fleets (Weavaway Travel of Newbury, Lloyds Coaches of Machynlleth and Konectbus of Norwich, for example).

Nationally, the Stagecoach Group is one of the largest users of both the diesel and hybrid versions, although a significant proportion of their non-hybrid Enviro 400s have been built onto Scania's N230UD chassis. They have allocated their 400H hybrids to busy city centre services including Oxford, Manchester, Newcastle and Sheffield, where they carry a green version of the corporate Stagecoach bus livery, complete with side panel wording "electric hybrid – for a cleaner greener city".

Early international interest has been shown in the Alexander Dennis Enviro 400 Trident, with a pair of dual-door, 10.5 metre air- conditioned buses being shipped to Hong Kong in 2009 for evaluation and trials. One was for Hong Kong Citybus (PC6795, fleet number 7000) and the other for Kowloon Motor Bus (PC4053, fleet number ATES1). Having been used to three axle double deck buses for many years, the Hong Kong Transport Department needed to give their permission for these vehicles to be used, as their weight exceeded the maximum permitted for two axle buses. Citybus received a second bus (7001, PH4891) in June 2010, with local observers suggesting that a larger order may be forthcoming in the future.

In 2011, Alexander Dennis introduced a low height open top Enviro 400 "tourist bus" to its range, produced in partnership with the Spanish bodybuilder UNVI, and aimed at the North American market and other countries subject to vehicle height or low bridge restrictions. Powered by a Cummins 8.9-litre engine and a four-speed automatic transmission, the "Urbis" low floor bus has a typical seating configuration of 51 on the upper deck and 29 (or 22 and 2 wheelchairs) on the lower.

The second generation of Enviro 400s was unveiled on 1st May 2014 as the "Enviro 400 MMC" (Major Model Change), incorporating Euro-6 engines and restyled bodywork as standard, with the Go-Ahead and Stagecoach groups, Oxford Bus Company and Reading Buses amongst the first operators. Deviating from UK manufacturers, the standard chassis option is Scania's N230UD/N250UD, with a gas-powered option available on the N280UD chassis any electric-hybrid options on Volvo's B5LH chassis. Unveiled in October 2015, Alexander Dennis introduced the Enviro 400H City as an integral hybrid option, which competes with (and has similar styling to) the New Bus for London. Arriva London was the first operator on its 78 service between Shoreditch and Nunhead, with the buses being designated AH-class. Blackpool Transport introduced high-specification examples in January 2016 for service 9 between Blackpool and Clevelys.

Alexander Dennis Enviro 200 Dart & 200H (2006-Current)

Above: The Alexander Dennis Enviro 200 Dart has become a popular single deck bus in use within London. LX58CAE was delivered in 2008 and now operates for Stagecoach London's Selkent fleet.

The second generation of Enviro 200 Dart was designed to be the successor to the Dennis Dart SLF and the unsuccessful first generation of Enviro 200 (only five being built between 2003 and 2007) and was unveiled by Alexander Dennis at their Guildford factory in August 2006. The low-floor chassis and body combination is built by Alexander Dennis, but the chassis was also made available separately for completion with, for example, East Lancs Esteem and MCV Evolution bus bodywork.

Alongside other members of the Enviro family, the chassis and mechanical units are constructed at the Alexander Dennis factory in Guildford, and then transported by lorry to the Plaxton factory at Scarborough for the addition of bodywork. Like the Dart SLF, the Enviro 200 Dart is available in different lengths from 8.9 to 11.3 metre and is 2.44 metres wide. Alexander Dennis versions are powered by Cummins four or six-cylinder Euro 4 engines (later vehicles being fitted with Euro 5 engines) attached to either Voith or Allison transmission options. The bodywork has been designed to maximise seating capacity, which varies between 26 and 39 passengers, dependent on vehicle layout, and is available in both single and dual door format. The aluminium body panels and glazing are bonded, and the front and rear panels are both GRP mouldings. The front panel styling is similar to that used on the single-deck Enviro 300 and double-deck Enviro 400, emphasising a consistent theme across the Enviro range.

From 2007, the Enviro 200 Dart body was made available on MAN's 14.240 chassis, including MAN's D0836 engine, Voith or ZF gearboxes and "exhaust gas recirculation" technology. 2008 saw the introduction of a hybrid-electric version, known as the Enviro 200H. Following research by Alexander Dennis and BAE Systems, the Enviro 200H is fitted with a four-cylinder Cummins ISBe 4.5-litre engine and a BAE Systems HybriDrive driveline capable of 140kW power output. The hybrid model reduces both fuel consumption and greenhouse gas emissions by over 30%. The 200H's lithium ion batteries do not require mains charging during their normal service life, with the vehicle's regenerative braking channelling energy back into them from normal driving. Whilst the hybrid-electric 200H is more expensive than the standard diesel-powered Enviro 200, many operators have taken advantage of the Government's Green Bus Fund scheme which contributes towards the price difference. In 2011, further modifications were made to ensure that the Enviro 200 Dart could meet the European Community Whole Vehicle Type Approval.

Above: Southdown PSV is a small bus operator based in Copthorne, West Sussex. YX59BZD is one of three Alexander Dennis Enviro 200s delivered new in 2009 for tendered services 236 (between East Grinstead and Oxted) and 494 (Oxted to Caterham-on-the-Hill). These Enviro 200s joined two examples already at work in their fleet.

The third generation of the Enviro 200 was announced at the Euro Bus Expo at Birmingham's NEC as the "Enviro 200 MMC" (Major Model Change) in November 2014, with National Express West Midlands taking the first production example (YX15OYO) in May 2015. This model also replaced the previous Enviro 300. Whilst conventionally powered 200 MMC buses can be seen across the country, in 2015 Alexander Dennis opted to provide the 200 MMC bodywork for completing a batch of Chinese BYD K9 electric buses, initially for Go-Ahead London General "Red Arrow" services 507 and 521 in Central London. London's Metroline followed soon after, and further batches for major operators are now in service.

From 2007, the Enviro 200 Dart body was made available on MAN's 14.240 chassis, including MAN's D0836 engine, Voith or ZF gearboxes and "exhaust gas recirculation" technology. 2008 saw the introduction of a hybrid-electric version, known as the Enviro 200H. Following research by Alexander Dennis and BAE Systems, the Enviro 200H is fitted with a four-cylinder Cummins ISBe 4.5-litre engine and a BAE Systems HybriDrive driveline capable of 140kW power output. The hybrid model reduces both fuel consumption and greenhouse gas emissions by over 30%. The 200H's lithium ion batteries do not require mains charging during their normal service life, with the vehicle's regenerative braking channelling energy back into them from normal driving. Whilst the hybrid-electric 200H is more expensive than the standard diesel-powered Enviro 200, many operators have taken advantage of the Government's Green Bus Fund scheme which contributes towards the price difference. In 2011, further modifications were made to ensure that the Enviro 200 Dart could meet the European Community Whole Vehicle Type Approval.

The third generation of the Enviro 200 was announced at the Euro Bus Expo at Birmingham's NEC as the "Enviro 200 MMC" (Major Model Change) in November 2014, with National Express West Midlands taking the first production example (YX15OYO) in May 2015. This model also replaced the previous Enviro 300. Whilst conventionally powered 200 MMC buses can be seen across the country, in 2015 Alexander Dennis opted to provide the 200 MMC bodywork for completing a batch of Chinese BYD K9 electric buses, initially for Go-Ahead London General "Red Arrow" services 507 and 521 in Central London. London's Metroline followed soon after, and further batches for major operators are now in service.

Optare Versa (2007-Current)

Optare unveiled their new Versa midibus design at the Euro Bus Expo at the NEC in Birmingham in November 2006. The Versa had been designed to fill a void in Optare's existing product range, being larger than the Optare Solo minibus and smaller than the full-size Optare Tempo. The Versa is available as a 9.7, 10.4, 11.1 or 11.7 metre integrally constructed bus, is 2.5m wide, and seats 36 to 44 passengers, dependent upon the chassis length selected. A 12-metre school bus version is also available which can seat 57 passengers.

Above: Safeguard of Guildford is a loyal operator of Optare buses, mostly allocated to Guildford circular services to the nearby Park Barn Estate. MX58ABF was the first Versa to be operated, being delivered in a plain white livery in 2008, and at the time joining Optare Excels and a single Optare Tempo within Safeguard's smart bus fleet.

The Versa incorporates a small wheelbase, just over five metres on the shortest version, which greatly assists the negotiation of tight turns. This addresses one of the concerns from operators of the long wheelbase Optare Solo, which with a forward front axle and seven-metre wheelbase suffered from poor manoeuvrability. The Versa is powered by Euro 4 specification engines provided by Cummins, MAN or Mercedes, which deliver drive through an Allison automatic transmission. The "power pack" is cradle mounted which allows easier removal and replacement when maintenance is necessary. Within the driver's cab, Optare's EcoDrive dashboard option encourages responsible and environmentally friendly driving. The Versa's modern external styling includes an aerodynamic front profile, curved glazing and the use of LED lighting at both the front and rear.

The first deliveries of the Versa commenced in 2007, including an order for 25 placed by Stagecoach Western on its launch day a year earlier. Other early adopters included Arriva Midlands, Arriva Shires & Essex, Transdev Burnley & Pendle, Transdev Lancashire United and Transdev Yellow Buses in Bournemouth. 2008 saw the first requirement for two door Versas, with an order for nineteen 10.4m buses to Transport for London specification for Transdev London United's route 391 between Richmond and Sands End. More recently the Versa has joined the fleets of Go North East, East London, NCP Challenger, Nottingham City Transport and Thamesdown Transport, as well as a growing number of smaller independent operators such as Anglian Bus of Beccles and Safeguard of Guildford.

Optare have developed several environmentally friendly power options for the Versa, including full electric, Siemens diesel-electric hybrid, bio-methane dual fuel, and a unit that runs on plant oil or used cooking oil fuels. The Government's Green Bus Fund assisted in the purchase of diesel hybrid Optare Versas for some operators: those having taken advantage so far include Greater Manchester PTE/First Manchester, including the use of hybrid Versas on the Manchester Metroshuttle service from 2010, Johnsons of Henley-in-Arden who use five hybrids on the Stratford-upon-Avon Park and Ride service, Rotala plc who ordered fifteen for services in Birmingham services and six that have been placed in service by Manchester Airport Group.

Because of working capital challenges, Optare agreed to sell nearly half of its shareholding to the Indian Hinduja brothers, the owners of the Indian Ashok Leyland Company, in January 2012 taking their shareholding in Optare up to 75%. Some Optare buses have now started to be offered to the Indian market, such as the Ashok Leyland Solo (see Optare Solo) and are notable for carrying the Leyland "wheel" badge last seen in the UK several decades ago.

Optare Rapta (2008)

Above: Unveiled at the 2008 Euro Bus Expo at the NEC,Birmingham, Optare's only "Rapta" integral double deck bus was not well received by the industry. It was dismantled in mid-2009, with Optare choosing to provide their own chassis for their Optare Olympus body instead.

Optare unveiled their new Rapta integral double deck bus at the Euro Bus Expo at the NEC in Birmingham in November 2008. It was offered at varying lengths between 10.1 and 11.9 metres and was environmentally friendly, with either a MAN Euro 5 compliant engine with EGR (exhaust gas recirculation) or an optional Enova lithium ion hybrid system. It was a low floor design, offering a seated capacity of between 70 and 86 passengers dependent upon individual configuration, and shared new styling similarities with the new Optare Solo + which was on display at the same event. It was fitted with two large tree deflectors on both front corners, to protect the large areas of glass. Industry feedback on the Rapta was not favourable and plans for its introduction were dropped by the middle of 2009, with the sole Rapta having been dismantled by the end of the year.

Wrightbus StreetLite (2010-Current)

Based in Ballymena, Northern Ireland, Robert Wright & Son Coachbuilders built their first bus bodywork in 1978. Mainstream production commenced with their "Handybus" body for the Dennis Dart in the early 1990s, and later the Endurance body for Volvo B10B and Scania N113CRB buses. The low-floor Pathfinder was launched in 1993 for the Dennis Dart SLF and Scania N113CRL, followed by the Liberator in 1995 (for the Volvo B10L chassis) and the Renown (for the Volvo B10BLE).

Introduced in 2010, the Wrightbus StreetLite is a low floor midibus, initially only offered in "wheel forward" format, with the front axle ahead of the entrance door. The additional "door forward" design, with the front axle being located behind the entrance door, was added in 2011, and the longer StreetLite Max a year later in 2012. The majority of StreetLites are powered by diesel engines (most recently either Daimler or Cummins Euro-6 units) coupled to Voith six-speed automatic gearboxes. 2013 marked the introduction of the "Micro Hybrid" option, which whilst not being a full hybrid, uses regenerative energy capture from vehicle braking to power the bus electric and compressed air systems, saving up to 10% on diesel costs. Since 2018 all models have been made available as hybrid-electric or full-electric powered – First South Yorkshire and Stagecoach Yorkshire taking the first hybrid

StreetLite Max variants in early 2018. Arriva Shires & Essex has also operated a batch of hybrid StreetLite WF buses from its Milton Keynes, Luton and High Wycombe depots.

Above: SK18TLY is a 2018 Wright StreetLite DF in the fleet of Hallmark Buses (part of the Rotala Group). It is seen in Staines bus station at the end of a 458 working from Kingston-upon-Thames.

The Wright StreetAir introduced in August 2016 (and replacing the StreetLite EV) provides for three methods of recharging the bus batteries. The first is conventional overnight charging using a plug-in system, the second is an induction capability through the road surface and the third is a conductive option using pantographs to an overhead supply.

Over 1,600 StreetLites are in service within the UK, with an increasing amount working on Transport for London services with operators such as London General, Quality Line of Epsom and Tower Transit. The First

Group own and operate over 500 of the type. They are perhaps most widely recognised as the buses which supported the London 2012 Torch Relay as it made its way around the country, with ten StreetLites from Stagecoach South Wales carrying a special orange livery for this memorable duty.

Wrightbus StreetLite Specifications

- StreetLite WF "Wheel Forward", 8.8 metres, 33 passengers
- StreetLite WF "Wheel Forward", 9.5 metres, 37 passengers
- StreetLite DF "Door Forward", 10.2 metres, 37 passengers
- StreetLite DF "Door Forward", 10.8 metres, 41 passengers
- StreetLite Max "Door Forward", 11.5 metres, 45 passengers
- StreetLite EV – battery electric bus, available in all lengths (to 2016)
- StreetAir EV WF – battery electric "Wheel Forward" (from 2016)
- StreetAir EV DF – battery electric "Door Forward" (from 2016)

Wrightbus "New Routemaster" (2011-2017)

The New Routemaster (originally "New Bus for London" or NB4L) is a modern replacement for London's iconic AEC Routemaster from the 1950s and 6-s. With the original London Routemasters having been retired by London Mayor Ken Livingstone in 2005 in favour or articulated Mercedes Citaro "bendy-buses", Boris Johnson made buses an London Mayoral Election issue in 2008 and upon being declared Mayor set about delivering his promise to develop and launch a "New Bus for London". By the end of the year, a competition to design the new bus had attracted over 700 responses, and at the beginning of 2009, six manufacturers submitted tenders to design and build it. Ultimately Wrightbus of Ballymena in Northern Ireland was declared the successful company, and design work commenced in early 2010. Wrightbus worked with Transport for London and the Heatherwick Studio to produce the styling for the bus, which was to retain and enhance much of the styling and functionality of the original Routemaster.

Above: New Routemaster LT669 (LTZ1669) from the Abellio London fleet, crosses Waterloo Bridge working a 68 service to West Norwood.

The Wrightbus contract required a bus that with a capacity of at least 87 persons, incorporating two staircases and three sets of doors to allow for the rapid boarding and exit of passengers. The rear most doors surround an open platform, which was anticipated to be open during the day and closed off when not required, for example on night services. In common with many recent new bus deliveries to the Capital, the New Bus for London is powered by a diesel electric hybrid drive, using a Siemens electric drive and a Cummins ISBe 4.5-litre engine, and is claimed to be significantly more fuel efficient than other hybrid buses already in use. LT1-LT272 are Euro-5, with all others being Euro-6. The styling of the new bus incorporates an unconventional diagonal glazing style for both the front and rear staircases, so that they are lit by natural light. The rear staircase is in a similar position to the original Routemaster, and the front staircase situated just behind the driver. Passengers benefit from air conditioning, LED lighting, WiFi internet connectivity and full wheelchair accessibility. Transport for London has applied for Registered Design Protection, as it has been considered to be an iconic bus for London.

A static full-size model was displayed at the London Transport Museum's Action Depot on 11th November 2010, six months prior to London Mayor Boris Johnson taking the wheel of the engineering prototype at Wrights premises in Ballymena on 27th May 2011. The prototype underwent extensive trials at the Millbrook testing ground in Bedfordshire. The first complete NB4L, fleet number LT1 (LT61AHT), was unveiled to the public in London in December 2011 and driven from City Hall in Westminster to Trafalgar Square. It then commenced an extended tour of many London boroughs over the following weeks, allowing the general public to see the "New Bus for London" design for themselves.

The first operational NB4L (fleet number LT2, LT61BHT) entered service on Arriva London's route 38 between Victoria Station and Clapton on 27th February 2012, with seven more joining it on the route (LT1-LT8). Originally, when running with an open rear platform during daylight operating hours, a "Passenger Assistant" was employed, although they were not required to collect fare and were primarily intended to provide customer information and manage the safety of passengers. From LT9 (LTZ1009) onwards, the New Routemasters received registrations from the Northern Ireland block LTZ1xxx (and later LTZ2xxx), and sometime later LT1-LT8 were re-registered into the same series.

They have not been without occasional controversy - the poor air conditioning and lack of opening windows on early examples lead to a major re-glazing programme. Staff reductions have removed the Passenger Assistant from the rear platform, and buses now operate as OMO (one-man operated) with the rear platform doors controlled by the driver. The NB4L fleet always has a percentage of its number carrying colourful all-over advertising schemes (most commonly for mobile phones, sportswear and fashion brands) – these are vinyl wraps applied to the bus which allows for them to be frequently updated. Subsequently 1,000 examples have been produced and introduced into service, with the NB4L becoming the main vehicle type for many London routes – the last being LT1000 (LTZ2100) for RATP London United, which can normally be found on route 267 between Hammersmith and Fulwell.

Above: at any one time, a number of New Routemasters carry all-over advertising schemes. Go-Ahead London Central LT455 (LTZ1455) rounds Trafalgar Square, carrying an appropriate scheme for a double-decker bus as it heads for Dulwich Library on route 12.

Whilst a number of operators outside of the Capital have trialled the New Routemaster (including LT2 with First West Yorkshire and LT312 and 313 with Stagecoach Strathtay in Dundee), it currently can only be seen at work in London.

Following completion of the New Routemasters, Wrightbus has continued to develop and supply the Wright SRM, which takes the New

Routemaster style of bodywork and fits it onto Volvo's B5LH and B5LHC chassis, albeit without a second staircase and with only two sets of doors. The first SRMs were supplied to London Sovereign in September 2016 for route 13 (North Finchley to Victoria Station).

One anomaly to note is "ST812" (LTZ1812), which entered service with Metroline on route 91 on 9th May 2016, was built as a short version and is only 10.1 metres long. It seats eight less passengers than a standard New Routemaster, being H36/18T.

Wrightbus StreetDeck (2014-Current)

Northern Ireland-based Wrightbus unveiled a new integral double-deck buys design in 2014 – the "StreetDeck", powered by a 5.1-litre four-cylinder Daimler OM934 engine (Euro-6), coupled to a Voith four-speed or ZF six-speed automatic gearbox. They are typically configured to seat 73 with space for 27 standees, although larger capacity configurations are offered.

Early prototypes were made available towards the end of 2014, and were evaluated by Arriva Derby, Arriva London, First Greater Manchester, First South Yorkshire, Go North East and London Central. Go-Ahead Group subsidiary Brighton & Hove announced a multi-million-pound investment in January 2015 by ordering 24 StreetDecks with mobile phone charging points and free wi-fi connectivity to update the existing "Coaster" group of services from Brighton to Peacehaven, Newhaven, Seaford, Eastbourne and the South Downs National Park.

Wrights additionally offered the StreetDeck as a "Micro Hybrid" option, which whilst not being a full hybrid, uses regenerative energy capture from vehicle braking to power the bus electric and compressed air systems, saving up to 10% on diesel costs. Reading Buses placed a £1.3m order for an initial six of the type in March 2016, which operate "orange route" services 13 and 14 between Reading and Woodley and are the first buses for Reading which benefit from engine start-stop technology. From the Go-Ahead Group, Oxford Bus introduced 30 low-height (13ft 10in) Micro Hybrid StreetDecks in 2017, with twenty for the city's busy park and ride services. Recognising the large student population resident in and around Oxford, mobile phone charging has been supplemented by a leather-trimmed corner lounge area just behind the staircase!

Above: the third batch of Wright StreetDecks for the Oxford Bus Company were destined for route 5 and carried a distinctive yellow scheme for the route. Passing through Oxford City Centre, 683 (SK17HHN) is a low height example of the StreetDeck.

Hybrid-electric versions were made available in 2018, and once again First South Yorkshire was an early customer in June of that year, including converting Leeds' Elland Road and Temple Green park and ride services to StreetDeck operation. According to First, trials of the type demonstrated a 34% reduction in carbon dioxide emissions when compared with a Euro-5 diesel bus.

The StreetDeck has also been used by international operators, including a single demonstrator evaluated by each of Kowloon Motor Bus in Hong Kong, and Buses Vule in Chile. The Hong Kong trial resulted in an order for fifty StreetDecks for KMB being placed in January 2019, these requiring the six-cylinder Daimler engine to power the air conditioning required for the climate. Additionally, five Micro Hybrids have been delivered to the Nuevo Leon region of Mexico.

Glossary

AEC

The Associated Equipment Company of London produced commercial vehicles including buses from 1912 until their closure in 1979. It was created as part of the Underground Group's takeover of the London General Omnibus Company in 1912 and allowed all bus manufacturing activities into a separate division. In 1927, AEC moved from Walthamstow to new premises in Southall, and 1929 saw the introduction of a new range of bus chassis brands, all of which started with the letter "R" (as opposed to "M" for lorry chassis). AEC worked with English Electric to produce trolleybuses from 1931 until the start of the Second World War, and from 1941 all production was diverted to commercial vehicles to support the war effort. After the war, AEC and Leyland Motors combined to form British United Traction (BUT) to continue the production of trolleybuses and associated equipment.

Several takeovers commenced in 1948, with AEC acquiring Crossley Motors and the Maudslay Motor Company, and soon after changing its own name to Associated Commercial Vehicles ("ACV"), although the AEC brand survived until the very end of the company. ACV purchased the bodybuilder Park Royal Vehicles in 1949, together with its Leeds based Charles H Roe subsidiary. Leyland Motors took over AEC in 1962, and the final AEC double deck chassis for the Regent V and Routemaster were completed in 1968. The single deck AEC Swift and Reliance continued until 1975 and 1979 respectively.

BACo

The British Aluminium Company Limited was formed in 1894 and produced aluminium by bauxite ore extraction. It was based in Scotland where it used electricity produced by a number of hydro-electric dams and had a number of rolling mills in England. BACo produced bus bodywork for export orders, and several Hong Kong and Singaporean fleets ordered BACo bodywork, which had a characteristically angular design.

BMMO

The Birmingham and Midland Motor Omnibus Company was established in 1904 by a group of Midlands businessmen, but having failed to attract sufficient investment, their operation was acquired by British Electric Traction in 1905. Its buses carried the fleet name "Midland", and the operation soon became unofficially known as Midland Red after the colour of the livery they carried. After the First World War, Midland Red commenced express coach services. From 1923 until 1969, BMMO designed and built its own buses, initially known as "SOS" (generally accepted as meaning "Superior Omnibus Specification") and later "BMMO", and a limited number were supplied to other companies in the BET Group. Midland Red became part of the National Bus Company in 1969, and in 1974, BMMO was renamed to the Midland Red Omnibus Company following a transfer of assets and services to the West Midlands Passenger Transport Executive.

British Electric Traction (BET)

The British Electric Traction Company ("BET") was founded in 1895 and was a significant contributor to the electrification of tramways in Britain, Australia and New Zealand, as well as manufacturing trams from 1901. With municipal fleets commencing the compulsory acquisition of their own tramways, BET formed a bus operating subsidiary in 1905, which grew steadily up until the start of the Second World War. After the war, BET and the Tilling Group were the two largest bus operators in the country. BET sold its bus operations to the Government owned Transport Holding Company in late 1967, and these ultimately became part of the newly formed National Bus Company which became operational on 1st January 1969.

British Transport Commission (BTC)

The British Transport Commission ("BTC") was formed as a result of the Transport Act of 1947 to manage the country's railways, canals and road freight transport. The BTC's largest operations were in London, where it operated central area red buses, country area green buses and Green Line coaches. The Tilling Group sold its bus interests to the BTC in September 1948, soon followed by the Red and White Group two years later. Midland General also joined the group having previously been owned by the British Electricity Authority. The BTC also owned Bristol Tramways/Bristol Commercial Vehicles and Eastern Coach Works, which was a major influence on the vehicles supplied to BTC subsidiary companies. By the late 1950s the BTC was facing financial difficulties and was ultimately abolished by the Transport Act of 1962. From 1st January 1963, all BTC bus and coach operations passed to the newly formed Transport Holding Company, except for London's buses and the London Underground, which became part of the London Transport Board.

Bus Preservation

When some buses or coaches reach the end of their revenue-earning lives, they may be saved from scrapping by individuals, groups or museums that have an interest in preserving transport history for future generations. This activity varies from amateur preservationists to professional restorers and full-time museums, and the age, construction methods and final condition of the withdrawn bus or coach will determine the amount of restoration work needed to return them to their original condition, often including the sourcing of period fittings and a repaint into the original operator's colours. Whilst older or unusual vehicles are keenly sought for preservation, the increase in the rate of fleet renewal has led to more modern examples entering preservation whilst they are still comparatively young. Some bus preservationists look overseas to locate rare former UK vehicles, with the recent withdrawal of Malta's historic route buses in July 2011 being one event that attracted considerable bus enthusiast activity and potential preservation candidates. Several Maltese buses have already been purchased and returned to the UK.

China Motor Bus (CMB)

Acknowledging that this publication is about British buses and coaches, China Motor Bus of Hong Kong is worthy or a brief mention as it has purchased a great number of British buses over the years. CMB commenced services on the Kowloon Peninsula in 1924, and soon started to be awarded exclusive franchises by the Hong Kong Government. CMB's franchise eventually ended in 1998, and over the intervening period an interesting and varied of British buses were operated, including the Bedford OB, Guy Arab, Dennis Loline, Leyland Titan PD3, Leyland Atlantean, Leyland Fleetline, MCW Metrobus, Dennis Dominator, Dennis Condor, Leyland Olympian and MCW Metrorider. After 1998, many of the surviving buses transferred to New World First Bus, which later shipped some double deck buses to Australia and the UK for sightseeing and school bus duties.

Cummins

An American corporation, the Cummins Engine Company was founded in Indiana in 1919. It currently sells diesel engines and related systems in nearly 200 countries through a large international distributor and dealer framework.

Duple

Duple's origins are in the post-World War One era, when bodies were designed for the Ford Model T chassis. A small facility was relocated to a factory in Hendon in 1926, and coachwork production started in earnest with an early customer being the Great Western Railway for its bus fleet. 1930 saw an order for 50 bodies for Green Line AEC Regals, and from then until the outbreak of the Second World War, most production was for Bedford chassis of various types, including the OB. Duple was chosen to build utility double deck buses during the war, and also developed the wartime OWB variant of the Bedford OB. The 1950s saw Duple losing market share, with Tilling Group companies standardising on Bristol chassis with ECW bodywork, although the company still took on additional capacity at Kegworth in 1952 (following their purchase of Nudd Brothers & Lockyer Limited) and Loughborough in 1955, these sites combining to become known as Duple (Midland). 1958 saw Willowbrook being purchased, although it retained its name, and in 1960 Duple purchased the Blackpool operations of Burlingham, which became known as Duple (Northern). Duple continued to align many of its individual bodywork designs to Bedford chassis developments during the 1960s, for example the Bella Vista for the

Bedford VAS and the Vega Major and Viceroy for the Bedford VAL70. Willowbrook was sold in 1971.

In 1972 Duple unveiled the new Dominant body as a competitor to Plaxton's Panorama Elite, and until the 1980s new versions were regularly introduced to remain competitive. As European coaches began to find a following with some UK operators, Duple introduced three new designs to compete – the Caribbean and Laser in 1982 and the Calypso in 1983 – although none of these attracted orders of any notable quantity. Duple was purchased by the Hestair Group in 1983, the owners of Dennis, and two years later in 1985 a new 300-series coach model was introduced (a bus model followed in 1987), followed by the integral Duple 425 which was best known for being operated by Alder Valley on Londonlink commuter services. In 1988, Dennis introduced the Dennis Dart, and Duple produced its "Dartline" bodywork based upon its existing 300-series bus. Hestair sold Dennis and Duple to a management buyout in 1988, although a continuing decline in coach sales lead to the closure of the Duple operation in 1989. Plaxton bought the rights to the 300-series and the Duple 425, whilst Carlyle acquired the rights to the Dartline body.

Duple Metsec

West Midlands based Duple Metsec acquired the business of Metal Sections in 1980. The business was engaged in the production of prefabricated metal sections for the construction of bus bodywork, which were mostly shipped overseas in kit form for final assembly by the operator concerned. Duple Metsec "R" series bodywork was for single deck chassis, whilst the "W" series was for double deck vehicles such as the Dennis Dragon/Condor. A later body version known as the DM5000 was constructed specifically for the Dennis Trident 3 chassis and was popular with Far East operators in Hong Kong and Singapore. Duple became part of the Hestair Group in 1983 and then a subsidiary of Trinity Holdings which closed Duple in 1989. Duple Metsec continued alone, later being acquired by the Mayflower Corporation, and DM5000 production finally ended after the creation of TransBus International in 2001.

Eastern Coach Works (ECW)

United Automobile Services commenced a coach building operation in Lowestoft in 1920, which passed to the Eastern Counties Omnibus Company in 1931. The bus body building operation was separated in July 1936 to form Eastern Coach Works ("ECW"). Operations were suspended from 1940 until the end of the Second World War, and in 1947 ECW became a nationalised company under the British Transport Commission, supplying bus bodywork for Bristol chassis for state owned bus operators. ECW passed to the state-owned Transport Holding Company in 1963, who sold a 25% share in ECW to Leyland Motors in 1965, which allowed ECW to start selling bodywork to private sector operators. 1969 saw ECW becoming a joint venture between the newly formed National Bus Company and British Leyland, an agreement which ended in 1982 with Leyland taking full control before ultimately closing the business in 1987 after 67 years of bodywork production.

East Lancashire Coachbuilders

A Lancashire-based bus bodywork builder, formed in Blackburn in 1934, East Lancashire was for many decades the principal bodywork manufacturer for municipal fleets. It was sold in 1988 to the Drawlane Group, who provided a period of stability by specifying East Lancs bodywork for Drawlane's various bus operators, including London Country South West and North Western Road Car. The East Lancs EL2000 body was made available on rebodied Leyland Tigers as early as 1989, and elements of its design were used by East Lancs within the Leyland National Greenway programme, which commenced in 1991 and was undertaken in partnership with London & Country at their Reigate premises. The EL2000 was replaced by the "Flyte" design in 1996.

The late 1990s saw further new designs from East Lancs, both for single deck buses (eg Spryte for the Dennis Dart) and double deck buses (eg Lolyne for the Dennis Trident and Vyking for the Volvo B7TL), with an upgraded range in 2001 using the "Myllenium" name (models becoming the Myllenium Lolyne, Myllenium Vyking etc). A range of bodies specifically to fit Scania chassis followed next, the OmniTown for the single deck N94UB and the OmniDekka for double deck N94UD, N230UD and N270UD chassis. 2006 saw three new designs, the Esteem for single deck buses, the Olympus for double deck, and the Visionaire open-top bodywork for sightseeing buses built on Volvo's B9TL chassis. East Lancs went into administration in 2007, being bought out by the Darwen Group to become "Darwen East Lancs". Darwen also purchased Optare in July 2008, which brought the two companies together under the Optare name, and the Optare Esteem, Optare Olympus and Optare Visionaire continued in production under new ownership until 2011, when they were retired in advance of a new Optare range expected in 2012.

Gardner

Gardner can trace its origins back to a small machinery manufacturer in Manchester in the 1860s but did not start producing diesel engines until 1903. Following a period as a munitions factory during the First World War, Gardner developed and launched their "LW" range of diesel engines for road vehicles, although they were also used to power Royal Navy submarines in the Second World War. Gardner engines remained a popular choice for buses from the 1950s until the 1980s, being trusted by many bus manufacturers, and it was changing emissions regulations in the 1990s that lead to them deciding to cease production.

Green Line

Green Line was created on 9th July 1930 by the London General Omnibus Company to operate a network of long-distance coach services from the Home Counties to Central London. Green Line became part of the London Passenger Transport Board on 1st July 1933, which also resulted in the absorption of all competing services within the London area. A growing network of cross-London services were developed until operations were suspended due to the Second World War, resuming in 1946. In January 1970, all Green Line services were transferred from London Transport to the new National Bus Company subsidiary London Country Bus Services, who continued to provide an extensive range of coach services until privatisation in 1986.

Whilst some of the privatised companies initially continued to market their coach services as Green Line, the name was ultimately transferred to Arriva, who continue to provide services from Hertfordshire and Bedfordshire to Central London and Heathrow Airport. The Green Line brand has previously been licensed by Arriva to New Enterprise Coaches and Stephensons of Essex and is currently used by Reading Buses who continue to operate services between Bracknell, Windsor, Slough and Central London.

Hybrid Bus

A bus which combines two power sources, a conventional internal combustion engine with an electric propulsion system. The most common combination is a diesel engined bus with a battery driven electric system, known as a "diesel electric hybrid". The two systems work together, in many cases the diesel engine is much smaller than the size needed to run a conventional bus, its main role in a hybrid being to provide power for the vehicle's air compressor, alternator and hydraulic systems. Modern hybrids use lithium ion batteries, which do not normally need external charging as they replace lost battery power using energy transferred from braking activities, which is known as "regenerative braking". Hybrid buses are amongst the most environmentally friendly vehicles in use today, not only providing a significant reduction in carbon emissions, but also providing a much quieter and smoother ride for passengers.

Integral Chassis

A bus or coach built as an entire unit, often by the same manufacturer, often without the use of a separate chassis or underframe. Examples within this publication include the Leyland National and Optare Versa.

Kowloon Motor Bus (KMB)

Acknowledging that this publication is about British buses and coaches, Kowloon Motor Bus of Hong Kong is worthy or a brief mention as it has purchased a great number of British buses over the years. KMB was created by the Hong Kong Government in 1933, and its first British purchases were 20 Daimlers in 1949. More recent imports have included Dennis Dominators, Dennis Dragons, Leyland Fleetlines, a single MCW Metrorider, three-axle MCW Super Metrobuses and Alexander Dennis Enviro 400s. KMB is part of Transport International Holdings Limited, operating over 4,000 buses on 399 bus services on the Kowloon Peninsula and in Hong Kong's New Territories.

London Country Bus Services (LCBS)

The Transport (London) Act of 1969 created London Country Bus Services from 1st January 1970 as a subsidiary company of the newly formed National Bus Company, and to take over the Country Area bus and Green Line coach operations from London Transport. Headquartered in Reigate, Surrey the London Country operating area completely encircled the Capital, with bus and coach routes in eight counties, provided using an ageing fleet of 1,267 vehicles transferred from London Transport. Much of the first ten years was spent on the task of modernising the fleet, with Leyland Atlanteans and Leyland Nationals becoming the mainstay of the bus fleet, and various types of AEC Reliance and Leyland Tiger updating the Green Line coach network. The 1985 Transport Act paved the way for bus deregulation, and London Country was split into four separate companies in September 1986 prior to their individual sales in 1988.

London Transport (LT)

The origin of bus transport in London go as far back as the horse drawn omnibus services provided by George Shillibeer from 1829, joined by the Thomas Tilling services in 1850 and the London General Omnibus Company (LGOC) in 1855. Horse drawn travel was replaced by the introduction of motor buses, in 1902 to LGOC and 1904 to Thomas Tilling, although the two rival operators soon agreed to co-operate by sharing their combined resources. LGOC was purchased by the Underground Group in 1912, and in 1933 the combined bus and underground operation were incorporated into a new organisation, the London Passenger Transport Board (LPTB). The LPTB became the London Transport Executive (LTE) in 1948, as part of the British Transport Commission, and following its abolition in 1963 was taken over by the London Transport Board until 1969. The Transport Act of 1968 split the Country Area (green buses) and Green Line coach services from the Central Area (red buses), with these passing to a new subsidiary of the National Bus Company, London Country Bus Services, on 1st January 1970. The Greater London Council operated London Transport's red buses until 1984, when London Regional Transport (LRT) assumed responsibility. The most recent change took place in 2000, when the Mayor of London's transport organisation Transport for London (TfL) took over. TfL currently operates as three divisions, with the red London buses being part of "Surface Transport", along with related activities such as London Dial-a-Ride services, Victoria Coach Station and CentreComm (London Buses' Command and Control Centre).

MCW (Metro Cammell Weymann)

Metro Cammell Weymann was formed in 1932 by the joining of Birmingham based Metro Cammell (the bus bodywork business of MCCW, the Metropolitan Cammell Carriage & Wagon Company) and Weymann Motor Bodies of London (and later Addlestone in Surrey). Bus bodywork was built at both locations until Weymann closed in 1966. From 1977, MCW started to produce its own chassis, including the MCW Metrobus bus, the MCW Metroliner coach and the MCW Metrorider midibus. In 1989, MCW's parent company chose to sell its bus and rail manufacturing companies, and as there was no interest in the group as a whole, MCW's individual products were sold separately. The Metrorider and Metroliner were purchased by Optare who soon relaunched the Optare MetroRider but chose not to progress the Metroliner. The double deck MCW Metrobus design passed to DAF (chassis) and Optare (bodywork), who collaborated to transform the vehicle into what later became known as the Optare Spectra.

National Bus Company (NBC)

Following the Transport Act of 1968, the National Bus Company ("NBC") was formed on 1st January 1969 to take ownership of a number of subsidiary bus operating companies within England and Wales, which at that time were operated by the Government's Transport

Holding Company (THC). The NBC expanded in 1970 to absorb the country bus operations of London Transport (becoming London Country Bus Services), Exeter (becoming part of Devon General), Luton (becoming part of United Counties) as well as the Gosport and Fareham Omnibus Company. NBC had inherited significant shareholdings in Bristol Commercial Vehicles and Eastern Coach Works upon its formation, and soon formed a partnership with British Leyland; these relationships were a major contribution to the choice of vehicles made available to NBC subsidiary companies. A corporate image followed in 1972 with the familiar "double N" logo, and compulsory fleet colours of either leaf green or poppy red on buses, and plain white on National Travel coaches. National Travel later became National Express, and a network of holiday services marketed as "National Holidays" was also implemented. The National Bus Company ceased in October 1986 at the point of UK bus services being deregulated and privatisation of its 52 subsidiary companies commenced. The "double N" logo was retained by National Express for continued use on its coach services, although has since disappeared.

National Express

The National Bus Company created the National Express brand in 1972 to unite all express bus and coach routes being provided by its subsidiary companies. Using an all-over white livery, services were initially branded "NATIONAL" using alternate blue and red letters and the familiar "double N" logo, with the "National Express" fleet names following shortly after. Almost every National Bus Company subsidiary operator contributed to the National Express network, with only their respective fleet name being added in small font to the otherwise standard coach livery. Some services in the 500-range were designated "Rapide" including an on-board service of drinks and snacks, and a toilet to reduce journey times by reducing the need for comfort breaks. Coach services were de-regulated in 1980, and the subsidiary bus operating companies privatised in 1986. National Express was floated on the London Stock Exchange in 1992, and since then has continued to provide a network of express coach services throughout the UK. Whilst National Express now operates some of its services directly, most routes are contracted or franchised to local bus and coach operators, and over 60 have become involved. A standard coach design by Salvador Caetano, the Levante, is provided on most modern-day services, being fitted to chassis manufactured by Scania (K-series) or Volvo (B9R, B11R or B12B).

Northern Counties

Based in the Lancashire town of Wigan, Northern Counties Motor and Engineering Company Limited was established in 1919, commencing bus bodywork construction in the early 1920s. Northern Counties was permitted to build utility buses during the Second World War, helping it to become supplier of mostly double deck bodywork to bus operators after the war. In 1967, it purchased the nearby body builder Massey Brothers Limited, based in Pemberton, retaining their premises to produce Northern Counties own bodywork. After the 1968 Transport Act created the combined "South East Lancashire North East Cheshire Passenger Transport Authority", more commonly referred to as SELNEC, Northern Counties worked with them to introduce a new bus body to standardise the variety in the different fleets which combined to make SELNEC. The 1972 Local Government Act added Wigan Corporation to SELNEC, creating the new Greater Manchester Public Transport Executive, and a large proportion of Northern Counties 1970s and 80s production was for the Manchester area. Quieter times followed in the 1980s and early 90s, and in 1995 Northern Counties was purchased by Henlys, the owners of Plaxton, who retired the Northern Counties name in 1999 in favour of their Plaxton brand.

Park Royal Vehicles (PRV)

Park Royal Vehicles of North West London has its origins in the nineteenth century, and owned a subsidiary company known as Roe (after its founder, Charles H. Roe) based in Leeds in West Yorkshire. In 1949, it was acquired by Associated Commercial Vehicles Ltd, who already owned AEC, and allowed the development and production of a variety of single and double deck designs, including the Park Royal bodied AEC Routemaster for London Transport. Associated Commercial Vehicles merged with Leyland Motors in 1962 to form the British Leyland Motor Corporation, which was itself nationalised in

1975. Park Royal was chosen to build the new Leyland Titan (B15) for London Transport, producing the first 250 chassis in 1978/9. Leyland was unhappy with Park Royal's production and announced that it was to be closed in November 1979: it subsequently set up its own facility to complete Leyland Titan production.

Plaxton

Based in the North Yorkshire seaside town of Scarborough, Plaxton commenced bodybuilding activities on Ford Model T chassis shortly after the First World War, although this was largely confined to small Yorkshire operators. By the 1930s, Plaxton was concentrating on the production of coach bodywork for chassis such as the Bedford WTB, although production ceased in 1939 when the Plaxton factory was acquired for munitions production. Production resumed in 1945 and grew steadily during the 1950s. In 1958, Sheffield United Tours requested Plaxton to design a new, more modern coach design, and this materialised as the "Panorama", characterised by large panoramic windows, straight body lines and the door at the front of the chassis ahead of the front axle. The Panorama underwent many changes over the next 16 years, including the introduction of a new "Panorama Elite" design at the 1968 London Commercial Motor Show, and it was eventually replaced in 1974 by the "Plaxton Supreme", which was a step forward in being of metal construction.

The "Plaxton Paramount" was unveiled at the 1982 British Motor Show, a new design incorporating a characteristic named and sloping window to the front of the bodywork. It was available in 8, 10, 11 and 12 metres versions (Paramount 3200), as well as high-floor (Paramount 3500) and a double deck coach (Paramount 4000), and modernised the coach industry during the 1980s. A final Paramount version, the Mk III, was introduced in 1986 and was used to create the "National Expressliner", which was used to further modernise National Express long distance coach services. Plaxton bought out its main competitor Duple for £4m in 1989. The Dennis Dart was introduced in 1989 and became an instant success, so Plaxton developed the Plaxton Pointer midibus body, which was launched in 1991, the same year as two new coach models, the Plaxton Premiere and Plaxton Excalibur. Plaxton purchased Northern Counties for £10m in 1995. Plaxton joined the TransBus International venture in 2000, which saw both Dennis and Plaxton names replaced with TransBus, although the operation failed in 2004, and Plaxton re-emerged independent once again in 2005. The final bodies introduced were the low floor Centro and Primo minibus in 2005, before Plaxton was itself acquired by Alexander Dennis in 2007. Under this new ownership, a brand new Plaxton Elite was unveiled in 2008, which continues to be made available for the Volvo B9R, B12B and B13R coach chassis.

Pneumocyclic Transmission

Urban and city centre buses start and stop frequently, which makes manual gearboxes unsuitable for this purpose. From the 1930s through to the 1960s, several manufacturers introduced "Preselector Transmissions", which allowed the driver to pre-select the next gear and then engage it when needed using a change gear pedal. The next progression in bus transmissions was the introduction of the new "Pneumocyclic Transmission", which combined the pre-select design and the change gear pedal into a single operation using a small lever mounted alongside the steering wheel, and which used air pressure to complete the gear change activity.

Red Arrow

London Transport introduced their "Red Arrow" network of limited stop, high frequency bus services in April 1966, initially using AEC Merlins. Following expansion of the network to eight services, the Merlins were replaced in 1981 by a batch of 69 Leyland National 2s, and 41 of them were treated to the Greenway refurbishment programme between 1992 and 1994. Privatisation in 1994 saw Red Arrow services passing to London General: the network was reduced to four services by 1998, and then further to two (507 and 521) by 2002. At this point the National 2s were replaced by 18-metre-long Mercedes Citaro O503G articulated buses which continued to carry the Red Arrow fleet name. "Bendy-buses" became a London Mayoral Election issue in 2008 and following Boris Johnson's victory these were soon replaced by London General with standard 12-metre-long Mercedes Citaro O530 buses in 2009, which provide seating for only 21 passengers with an additional 76 standees. In September 2016, the whole fleet was replaced with new Alexander Dennis Enviro 200-bodied BYD electric buses (SEe designation), the chassis being made in China.

Scottish Bus Group (SBG)

The Scottish Bus Group ("SBG") was a state-owned bus transport holding company covering all of Scotland. It originated from a 1961 decision to pass the Scottish bus operating subsidiaries of the British Transport Commission to a new organisation, which became known as Scottish Omnibuses Group Holdings, and in 1962 it became part of the Transport Holding Company. Its seven subsidiaries were Central SMT, Highland Omnibuses, Scottish Motor Traction, Walter Alexander & Sons (Fife, Midland and Northern) and Western SMT. It was formally renamed to the Scottish Bus Group in 1963, and ultimately became part of the Scottish Transport Group on 1st January 1969, the same date as the formation of the National Bus Company in England and Wales. A reorganisation in 1985 resulted in twelve operating companies and a separate engineering division, and these separate companies were all privatised at the start of the 1990s.

SELNEC

The Transport Act of 1968 established four large regional public transport executives ("PTE") including Merseyside PTE, Tyneside PTE and West Midlands PTE. The fourth PTE was established up for Manchester and its surrounding areas, and was known as SELNEC PTE, which stood for "South East Lancashire and North East Cheshire". Formed on 1st November 1969, SELNEC took over more than 2,500 buses that had previously been owned by the municipal fleets of Ashton, Bolton, Bury, Lancashire United Transport (in 1976), Leigh, Manchester, North Western (ex National Bus Company in 1972), Oldham, Ramsbottom, Rochdale, Salford, SHMD (Stalybridge, Hyde, Mossley and Dukinfield), Stockport and Wigan (in 1975). Greater Manchester became a metropolitan county on 1st April 1974, after which date SELNEC was renamed to Greater Manchester PTE.

Transport Holding Company (THC)

The Transport Holding Company ("THC") was created and owned by the British Government as a result of the 1962 Transport Act, which was implemented to controls a variety of state-owned companies previously managed by the British Transport Commission. The THC commenced operations on 1st January 1963 with 24 former BTC fleets in England and Wales, seven in Scotland, and shares in three others jointly owned by British Electric Traction (BET). It also took ownership of Bristol Commercial Vehicles and Eastern Coach Works. As the Transport Act of 1968 was passing through Parliament, BET sold all its bus operations to THC, which brought another 25 fleets (plus their respective subsidiaries) under THC control. THC was the basis upon which the National Bus Company (NBC) commenced operating on 1st January 1969.

Wallace Arnold

Established by Wallace Cunningham and Arnold Crowe in Leeds in 1912, Wallace Arnold was one of the UK's largest and most successful holiday coach tour specialists. With the deregulation of coach services in 1980, Wallace Arnold was instrumental in creating the British Coachways consortium of operators who wished to provide competing services against the larger state-owned National Express network, although it did not remain part of the consortium for long. Wallace Arnold merged with Shearings in 2005 to become WA Shearings, and in 2007 the "WA" was dropped to become Shearings Holidays.

Walter Alexander Coachbuilders

Walter Alexander commenced bus operations between Falkirk and Grangemouth in Scotland in 1913, and in 1924 established a company to manufacture buses to their own design. The formation of the Scottish Motor Traction Company (SMT) in 1928 allowed Walter Alexander access to a wide range of resources and sales opportunities and became the prominent bus bodywork provider for Scotland. Nationalisation of bus services in 1945 caused the creation of Walter Alexander& Co (Coachbuilders) Limited in 1948. In 1961, Alexander's combined bus operations were separated into Fife Scottish, Midland Scottish and Northern Scottish and a year later they joined the Scottish Bus Group. Alexander continued to build bus bodies at their Falkirk premises throughout the 1960s and expanded in 1969 by purchasing Northern Irish bodybuilder Potters to form Alexander (Belfast). A management buyout took place in 1992 and Alexander was then sold to the Mayflower Corporation in 1995, before becoming part of TransBus International in 2001. TransBus International failed in 2004, re-emerging as Alexander Dennis, although the Alexander Belfast was closed at this point.

Weymann

Weymann established its first bodywork factory in Putney, London in November 1925, moving to Addlestone in Surrey in 1928. It was successful in gaining contracts to build bodies for London General AEC Regals in 1930, and in 1932 Weymann formed a partnership with the Birmingham company Metro-Cammell which resulted in designs and skills being shared and started a period of steady growth until the Second World War. Of over 1,500 buses bodied during the war, over 700 of them were to wartime utility specification. Weymann built bodywork for over 2,000 of London Transport's AEC Regent III RTs between 1945 and 1953, followed by 500 bodies for "Green Goddess" Bedford fire engines for the Auxiliary Fire Service in 1954/5. Aside from more routine bodywork orders, Weymann built the bodywork for the third Routemaster prototype (RML3) in 1957, and the very last body for a British trolleybus onto a Sunbeam MF2B (301LJ) for Bournemouth Corporation in 1962. Metro-Cammell purchased Weymann in 1963, with the Addlestone factory closing in 1966.

Willowbrook

Willowbrook Limited was based in Derby Road, Loughborough and could trace its bodybuilding origins back to the 1920s. Willowbrook bodywork became a common sight before and after the Second World War and was frequently chosen by BET companies. Willowbrook bought the designs of Loughborough based Brush bodywork in 1952, when these became available for purchase, and subsequent advertising included the phrase "Incorporating Brush Patents". Willowbrook was purchased by Duple in 1958, the name being maintained, and from 1962 it was intended that the Willowbrook name would be applied to all heavyweight single and double deck chassis, with Duple's Midland brand being applied to vehicles built on lighter Bedford or Ford chassis. There are numerous examples where this did not happen.

Duple decided to sell Willowbrook in 1971, and Willowbrook then started to develop new models for the 1970s coach market. The "Spacecar 008" was launched at the 1974 Commercial Motor Show and resulted in reasonable orders for National Travel subsidiary companies, to be fitted to AEC Reliance, Bedford YMT/YRT and Leyland Leopard chassis until 1977. The next Willowbrook model, "003", started a political chapter, with the then current Prime Minister Jim Callaghan ordering the National Bus Company to place their entire dual-purpose coach requirements with Willowbrook following representation from Loughborough's MP. The first example was AFJ696T, which joined the Western National fleet in 1979, and from that point onwards the quality of the bodywork suffered as Willowbrook were unable to cope with the demand now placed upon them. Willowbrook continued to provide bus bodywork for UK and international orders during this period, including 200 double deck bodies for left hand drive Leyland Atlanteans for PTS, Baghdad (Iraq) in 1980. The continued and unacceptable deterioration of 003 coach body production forced many operators to cancel their orders with Willowbrook and move their chassis to Duple or Plaxton to be completed. Following legal action instigated by the National Bus Company, Willowbrook closed in 1983.

However, Willowbrook was revived once again in 1986, offering a range of bus and coach bodies, including the "Warrior" bus and "Crusader" coach that were suitable for fitting to older bus and coach chassis. These found some interest, mostly with smaller independent operators, although City of Oxford treated six Leyland Leopards to a new Warrior body in 1990/91, and Brighton Buses took two. Willowbrook finally ceased production in 1992.

- End -

www.ingramcontent.com/pod-product-compliance
Lightning Source LLC
Chambersburg PA
CBHW071432180526
45170CB00001B/308